QUANTITATIVE PROTEOME ANALYSIS

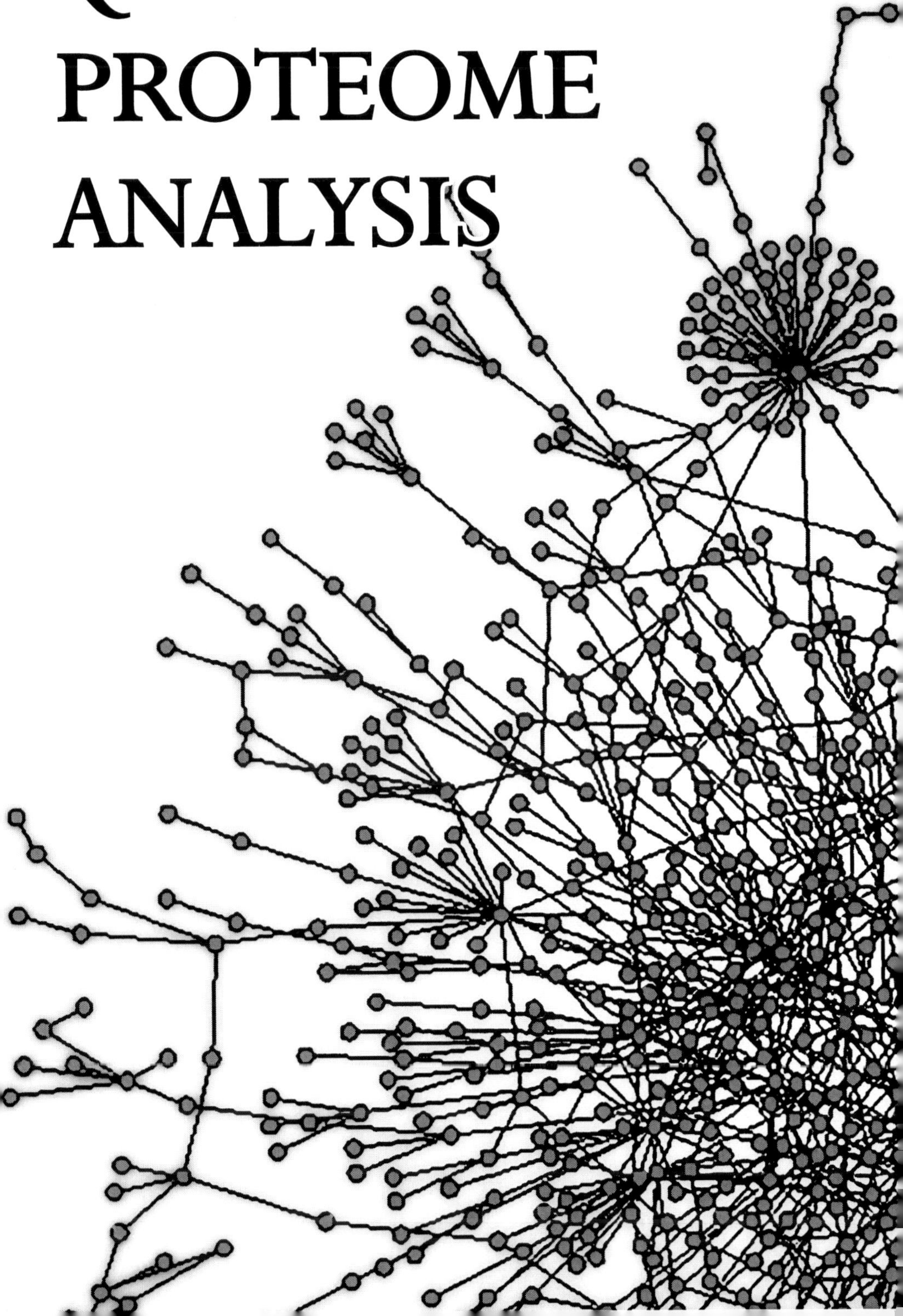

QUANTITATIVE PROTEOME ANALYSIS

METHODS AND APPLICATIONS

edited by

Kazuhiro Imai
Sam Y. F. Li

PAN STANFORD PUBLISHING

Published by

Pan Stanford Publishing Pte. Ltd.
Penthouse Level, Suntec Tower 3
8 Temasek Boulevard
Singapore 038988

Email: editorial@panstanford.com
Web: www.panstanford.com

British Library Cataloguing-in-Publication Data
A catalogue record for this book is available from the British Library.

Quantitative Proteome Analysis: Methods and Applications

ISBN 978-981-4316-51-4 (Hardcover)
ISBN 978-981-4316-98-9 (eBook)

Printed in the USA

Contents

Preface

Methods combined with separation techniques such as HPLC or electrophoresis for detection, identification, and quantification of proteins in complex biological matrices are essentially requisite to understand biochemical events in living cells, tissues, and organs. Especially, in searching for the difference in proteins, the specifically expressed proteins in various circumstances, such as between diseased and normal state, aging, exposed and unexposed by drugs, and so forth, a reproducible quantitative method to distinguish between minute changes in amounts of proteins is required. This book focuses on describing the advantages and disadvantages of each of such methods in terms of accuracy, sensitivity, and especially reproducibility. Its other key feature is the description of the effective application of these methods, which result in many discoveries regarding the role of the proteins expressed in living cells.

In the last decade, more than a thousand papers were published on the proteome analysis acquired by the application of these methods in various areas of life sciences. For junior researchers, the first major interest should be the selection of the method when they start their proteome analysis in valuable biological specimens. However, there are few textbooks or monographs showing the evaluation of each method. That is why this book is published.

Kazuhiro Imai

Summer 2013

Chapter 1

Fluorogenic Derivatization Followed by HPLC Quantification and Final Identification of Proteins by HPLC-Tandem Mass Spectrometry (FD-LC-MS/MS) Method

Akiyo Koshiyama, Tomoko Ichibangase, and Kazuhiro Imai
Research Institute of Pharmaceutical Sciences, Musashino University, 1-1-20 Shinmachi Nishitokyo-shi, Tokyo 202-8585, Japan
k-imai@musashino-u.ac.jp

1.1 Overview of the Method

Fluorometry is one of the extremely sensitive methods for detecting small amounts of biomolecules: hormones, amines, amino acids, peptides, proteins, carbohydrates, lipids, or nucleic acids. Thus, it should be most suitable for the detection, quantification and identification of the expressed proteins in proteomics studies. However, many of proteins do not fluoresce, so that fluorescent derivatization with reagents is necessary for the sensitive detection of proteins.

Quantitative Proteome Analysis: Methods and Applications
Edited by Kazuhiro Imai and Sam F. Y. Li

ISBN 978-981-4316-51-4 (Hardcover), 978-981-4316-98-9 (eBook)
www.panstanford.com

In this case, fluorogenic reagents that generate fluorescence after derivatization without fluorescence themselves are strongly recommended because of low background fluorescence affording highly sensitive detection of the derivatives. A few fluorogenic reagents for amines can be taken for the above use: orthophthalaldehyde [1] and fluorescamine [2]. As to the derivatization of cysteine residues in proteins, thiol specific reagents are requisites. Ammonium-7-fluoro-2,1,3-benzoxadiazole-4-sulfonate (SBD-F) [3] and 7-chloro-*N*-[2-(dimethylamino)ethyl]-2,1,3-benzoxadiazole-4-sulfonamide (DAABD-Cl) [4] are most appropriate among the existing reagents for the following reasons. First, their relatively small molecules enable them to react completely with cysteine residues in proteins under a mild condition. Second, derivatized proteins are hydrophilic and thus do not precipitate after the complete derivatization. Third, their derivatives are highly fluorescent with longer emission wavelength than natural fluorescence of biomolecule (less than 400 nm).

A chromatographic technique, high-performance liquid chromatography (HPLC) is also a powerful tool for separation of compounds in a complex matrix of bio-samples. However, it has not been usually used for protein separation in bio-samples except for the recently developed FD-LC-MS/MS method [3–6] for proteomics analysis. It consists of fluorogenic derivatization (FD) of proteins using fluorogenic reagent such as SBD-F or DAABD-Cl, followed by HPLC separation, detection and quantification of the derivatized proteins, isolation of the subject proteins, enzymatic digestion of the isolated proteins, and identification of the proteins utilizing HPLC and tandem MS with a database-searching algorithm. The schematic diagram of the FD-LC-MS/MS method is shown in Fig. 1.1. First, a proteins mixture is derivatized with a fluorogenic reagent and analyzed by HPLC-fluorescence detection. Next, a derivatized subject protein can be isolated without losing any amino acid sequence information including protein isoforms and post-translational modification, and then digested by trypsin. Finally, the obtained peptides mixture is subjected to HPLC, which is connected to electrospray ionization (ESI)-MS/MS, to identify the isolated protein using the probability-based protein identification algorithm. The derivatized proteins are detected and quantified at the femtomol level [3, 4]. The quantification is performed by calculating the peak intensity of the derivatized protein that corresponds to the amount of the derivatized protein existed. The accuracy of the method was acquired based on the reproducibility of the peak heights using the three representatives of the high, medium, and low peaks obtained from each individual mouse liver sample [7]: The relative standard deviation (RSD,%) for each between-day peak was less than 16 (a relatively high peak), 17 (medium peak), and 23% (a relatively low peak) (n = 3). The reproducibility of the retention time was also calculated using the relatively low peak, resulting in the between day RSD of 0.41% (n = 3).

A remarkable feature of the method is the use of the fluorescence for the sensitive detection [3, 4] and of HPLC for the reproducible quantitation of the derivatized proteins [7] as mentioned above. It also requires only simple apparatus, consisting of a pump, a column, and a fluorescence detector. Currently, a comprehensive profiling analysis in HPLC-fluorescence detection needs a 10 h operation. However, in case the elution time of a subject protein is determined, it would be possible to reduce the time required for an arbitrary analysis on the subject protein by re-optimizing the separation conditions. It would also be possible to reduce the overall analysis time by adopting a higher-performance column, such as a recently available non-porous reversed phase column. Moreover, any pre-treatments of the sample are not required in the procedure of the method since only fluorogenic derivatized proteins are detected without detection of non-fluorescent compounds on LC-chromatogram. In contrast, other proteomics analytical methods, such as two-dimensional polyacrylamide gel electrophoresis (2D-PAGE) and liquid chromatography-tandem mass spectrometry (LC-MS/MS: for example, ICAT and iTRAQ), require certain pre-treatments steps (precipitation, labeling, clean-up with column and enzymatic protein digestion: reviewed in [8]). In 2D-PAGE methods, pre-treatment procedure (precipitation etc.) for the sample is essential to obtain clear 2D electropherograms, because the composition of the sample is greatly influential to the resolution of every protein in electrophoresis. In case of ICAT and iTRAQ, each sample pre-treatment procedure is required since it influences the numbers of peptides observed by LC-MS analysis. However, such pre-treatment procedures would lower the sensitivity and reproducibility of the method owing to the loss of proteins existed in the sample during pre-treatment steps.

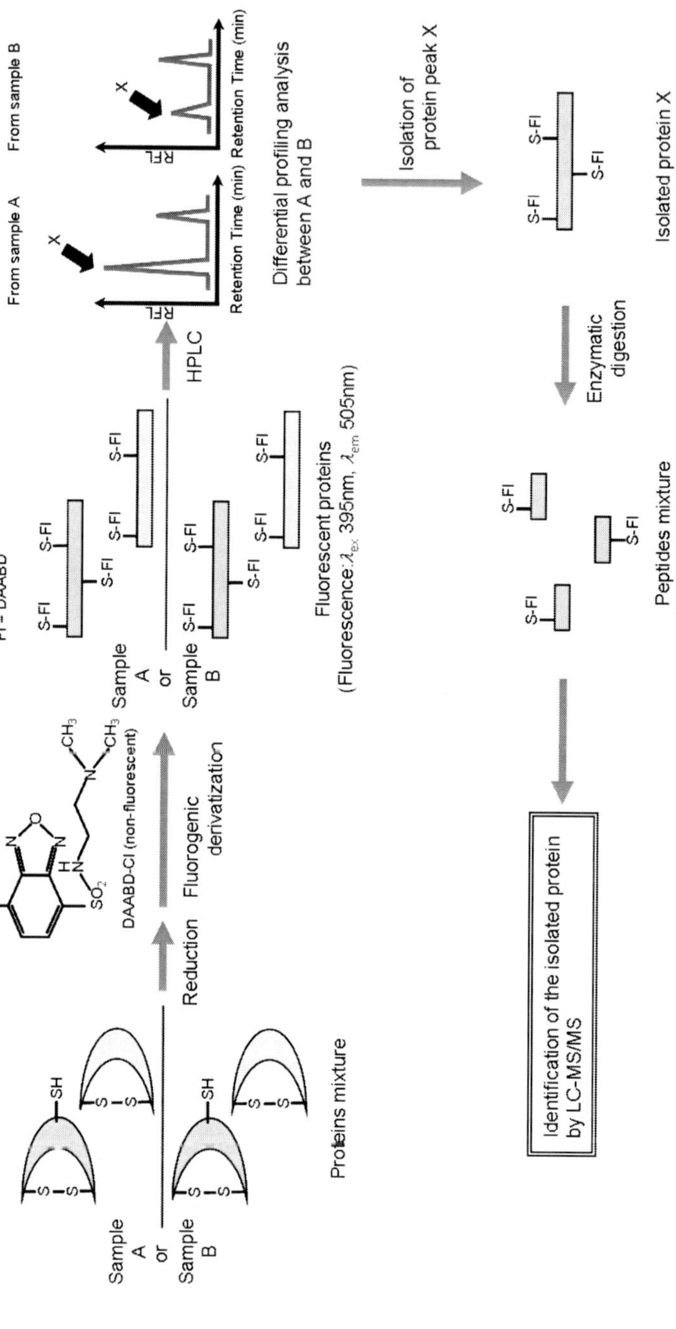

Figure 1.1 Flow diagram of the FD-LC-MS/MS method for differential proteomics analysis.

For a comparative analysis of the amount or species of proteins among two or more samples, namely a differential proteomics analysis, the FD-LC-MS/MS method is also useful. Actually, as shown in Fig. 1.1, a peak height on a LC-chromatogram of a subject sample can be compared with the corresponding peak height on that of the control sample in differential proteomics analyses (normal versus diseased cell, diseased versus treated cells etc.). Each fluctuating protein peak among samples is isolated and then identified by LC-MS/MS. The correspondence of the protein peak is judged from not only the specific retention time of the derivatives but also confirmation of the protein following isolation and identification of the proteins. As described in the text, the FD-LC-MS/MS method has been applicable to the extracts of mice liver [7], breast cancer cell lines [9], mice brain [10], Thoroughbred horse muscle [11], mice plasma [12], and *Caenorhabditis elegans* [13]. Further use of the method would reveal the role of certain proteins in the special biochemical events occurring in living animals.

1.2 Apparatus and Method

1.2.1 Apparatus

As described in Section 1.1, the FD-LC-MS/MS method does not require particular reagents, machines, or systems. In fluorogenic derivatization, all the necessary reagents including DAABD-Cl are available on the market. A HPLC system consists of a Hitachi L-7000 or L-2000 series and a fluorescence detector (Jasco FP-2025 plus or Hitachi L-2480). A protein column (Intrada WP-RP, 250 × 4.6 mm i.d., Imtakt Co.) is used for separation of the derivatized proteins. For identification of the derivatized protein that is isolated and tryptic digested, a nano-LC-ESI-tandem MS spectrometer (HCT plus, Bruker Daltonics) is used with the probability-based protein identification algorithm (MASCOT version 2.1.03, Matrix Science).

1.2.2 Method

Preparation of Samples and Determination of Total Proteins: Proteins are extracted as follows: Tissues are homogenized in 5 times volume of 10 mM CHAPS aq. with a Teflon pestle on ice. In case of the cell lysate, the lysate is dissolved in 10 mM CHAPS aq. and extracted with 20 times repeated pipeting on ice. The homogenate is centrifuged 20,400 g for 15 min at 4°C. The supernatant is then collected and stored as a soluble fraction at –20°C until use.

Fluorogenic Derivatization and HPLC conditions: Protein samples obtained from mice liver, breast cancer cells or Thoroughbred horse muscle are diluted with 10 mM CHAPS aq. to 4.0 mg/mL, while samples from mice brain were 20 mg/mL. Ten microliter of the sample were mixed with 60 μL of a mixture of 0.83 mM Tris (2-carboxyethyl) phosphine hydrochloride (TCEP), 3.3 mM ethylenediamine-*N,N,N′,N′*-tetraacetic acid sodium salt (Na_2EDTA), 16.6 mM CHAPS in the pH 8.7 buffer solution (6.0 M guanidine hydrochloride, Tokyo Chemical Industry), 25 μL of the buffer solution and 5.0 μL of 140 mM DAABD-Cl in acetonitrile. After the reaction mixture is placed on a 40°C water bath for 10 min, 3.0 μL of 20% trifluoroacetic acid (TFA) is added to stop the derivatization reaction. Twenty microliters of the reaction mixture (8.0 μg for mice liver, breast-cancer cells and horse muscle or 40 μg for mice brain) was injected into the HPLC system at a flow rate of 0.55 mL/min. As described above, the system consisted of a pump, a fluorescence detector and a protein column. Fluorescence detection was carried out at 395 and 505 nm of excitation and emission wavelengths, respectively. To further improve the column separation, column oven is set at 60°C. The mobile phase consists of acetonitrile, TFA and water with or without isopropanol. Protein mixtures are separated in 10 h. The mobile phase composition and gradient condition need to be optimized for each sample.

Identification of the derivatized proteins: Each eluate of the subject proteins is concentrated *in vacuo* and followed by the tryptic digestion. Twenty μL of peptide mixture is subjected to a nano-LC-ESI-tandem MS spectrometer (HCT plus, Bruker Daltonics) with an Ultimate/Famous/Switchos suite of instruments (LC Packings, Dionex). The MS/MS data is extracted as mgf files using Data Analysis 3.2 software (Bruker Daltonics). Peptides are identified by searching the mgf files against the National Center for Biotechnology Information (NCBI)-Cl, which picks the proteins including cysteine residues from the NCBI database using the MASCOT version 2.1.03 search engine.

1.3 Applications of the Method

Among the applications of the FD-LC-MS/MS method to the extracts of mice liver [7], breast cancer cell lines [9], mice brain [10], Thoroughbred horse muscle [11], mice plasma [12] and *Caenorhabditis elegans* [13], this review describes the results of the differential analyses concerning disease-related proteins in livers of hepatitis-infected mouse model [7], breast cancer-related proteins in breast cancer cells and normal epithelial cell [9], age-related proteins in small regions of mouse brain [10] and skeletal muscle proteins in fast Thoroughbred horses [11].

1.3.1 Differential Analysis of Disease-Related Proteins in Livers of Hepatitis-Infected Mouse Model

Hepatitis C virus (HCV) is the main cause of chronic hepatitis, which ultimately results in the progression of hepatocellular carcinoma (HCC) [7]. Moriya *et al.* have suggested that the HCV core protein plays a critical role in the progression of HCC and that the transgenic mice provide a good animal model for determining the molecular and pathological events in hepatocarcinogenesis with HCV infection [14–17]. Differential profiling analysis of proteins was performed using liver tissue from HCV core gene transgenic (Tg) and non-transgenic (NTg) mice by the FD-LC-MS/MS method. Tg/NTg ratios were calculated on the heights of the peaks corresponding to specific retention. Many proteins were up- or down-regulated during the progression of HCV-associated liver disease. Fifteen proteins were significantly altered in their levels of protein contents. The remarkable decrease in major urinary protein (MUP) and eukaryotic translation elongation factor 1 α 1 (EF-1α1) represents an early event in the progression of HCV-associated liver disease (at 6 months; early stages). MUP has been known as a negative tumor marker [18]. Suppression of EF-1α1 expression prevents the induction of apoptosis, with the regulation reflected in an antiapoptotic mode [19]. At 12 months of age (medium stages), proteins related to respiration, the electron transfer system, and defense against reactive oxygen species (ROS) were significantly up-regulated. Finally, at 16 months (late stages), proteins related to defense, oxidation and apoptosis significantly decreased. Cystatin B [20] and carbamoyl-phosphate synthetase 1 (CPS1) [21] are known to be down-regulated in tumor and/or carcinoma and exhibited a significant decrease with the present method. As a whole, the investigation of the differential expression of proteins in Tg and NTg mice revealed that many proteins related to biological functions such as respiration, protein synthesis, defense, oxidation and apoptosis fluctuate during the progression of chronic hepatitis C. Accordingly, a next step of proteomics investigation should be performed for a mitochondrial fraction, that is related to respiration, electron transport system, glycolytic system and apoptosis, and a study is now under investigation in our laboratory.

Anyway, these fluctuations may reflect a gross effect derived from the loss of liver function in the various stages of chronic hepatitis in HCV infection.

1.3.2 Differential Analysis of Breast Cancer-Related Proteins in Comparison among Breast Cancer Cells and a Normal Epithelial Cell

Breast cancer is a principal cause of death among women, mainly in old age. A few molecular markers for diagnosis and therapy have been found by mRNA profiling analysis [22, 23] and several proteomics studies [24, 25]. These include the milk mucins secreted by the tumor, the expressed estrogen receptor (ER), the human epidermal growth factor receptor type 2 (HER2) and a mucin-like carcinoma antigen CA15-3 [24]. The expressions of ER and HER2 are indicators, respectively, of treatment with the anti-estrogen drug, tamoxifen [25] and the monoclonal antibody, herceptin. However, not all breast-cancer cells expressed ER and a significant increase in the expression of HER2 was seen in only 25–30% of patients [26], thus limiting their use in therapy. In addition to that, CA15-3 was proposed as a marker for distant metastasis but was abandoned due to its poor reliability. Considering such background, other molecular markers for diagnosis and therapy of breast cancer are required. Differential proteomics analysis between breast cancer cells and a normal epithelial cell was performed using FD-LC-MS/MS method (Fig. 1.2) [9]. Seven breast cancer cell lines including metastatic cancer cell MDA-MB-231 were taken for investigation with a normal human mammary epithelial cell (HMEC) as a normal cell. To identify the differently expressed proteins, 13 proteins both universally expressed in the seven cancer cells and differently expressed from a normal cell were selected. This analysis demonstrated an exclusive expression of Ran binding protein-1 (RanBP-1) and peroxiredoxin-1 (Fig. 1.2) in cancer cells, with several up-regulated proteins, cofilin-1, Raf-1 kinase inhibitor protein (RKIP), profilin-1, Ras-related nuclear protein (Ran), thioredoxin-1, and Ras homology GDP dissociation inhibitor (RhoGDI). Also demonstrated were down-regulated proteins, stratifin, galectin-1, annexin-2, and elongation factor Tu (EF-Tu). Tropomyosin-1 was specifically expressed in a normal cell. In addition, among the seven cancer cells, differently expressed proteins were investigated in correlation with invasive and metastatic cancer cell MDA-MB-231 and with ER and HER2 receptors. Up-regulated RKIP in cancer cells would involve ER-positive, while down-regulated annexin-2, galectin-1 and EF-Tu would involve HER2-positive. The presence of cooperatively expressed annexin-2 and galectin-1 in a breast cancer cell without tropomyosin-1 was suggestive of the metastatic property. In conclusion, differential analysis of proteins using FD-LC-MS/MS method could propose a novel pathway correlated to invasive and metastatic breast cancer.

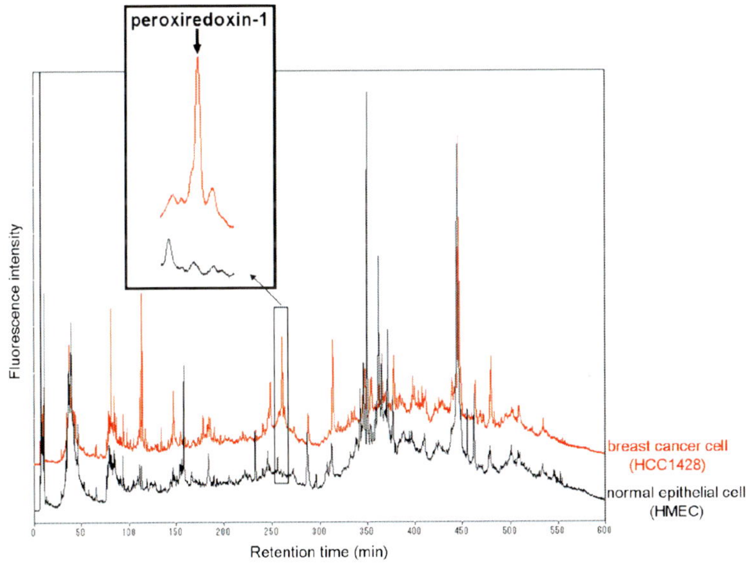

Figure 1.2 The chromatograms obtained from a breast cancer cell (HCC1428: red) and a normal epithelial cell (HMEC: black). The arrow shows a specific protein (peroxiredoxin-1) in the breast cancer cell.

1.3.3 Differential Analysis of Age-Related Proteins in Small Regions of Mouse Brain

In recent years, several laboratories attempted to identify the age-related proteins in mouse brain using the 2-DE method with whole brain as samples [27–31]. However, the brain functions differ among the regions and identification of age-related proteins in each region of brain is most important. The sensitivity acquired using the 2-DE method was insufficient to reveal the age-related proteins in each region. As a matter of fact, only three proteins were identified out of the 17 age-related proteins in mouse brain region detected by Tsugita *et al.* [32]. Thus, we investigated whether FD-LC-MS/MS method is applicable for identification of age-related proteins in small region of mouse brain [10]. Considering this background, differential proteomics analysis of aging in cerebral cortex (Ctx), hippocampus (Hippo) and brainstem (BS) of mouse brain was performed. We employed mice aged 4 weeks (childhood), 12 weeks (early adulthood) and 20 weeks (late adulthood). To reveal age-related altered proteins in brain regions (Ctx, Hippo and BS), the peak heights were compared with each other in the developmental stages (4, 12 and 20 weeks). As a result, 28 proteins were obtained as age-related altered proteins and were identified. Fifteen proteins were found to be altered significantly in Ctx and related to ribosome, cytoskeleton, synapses, glycolysis and antioxidant. In Hippo, 10 altered proteins were classified as ribosomal protein, cytoskeletal protein, synaptic protein, myelin protein and other proteins. In BS, three proteins were found in this analysis and classified as cytoskeleton protein, another class of proteins and enzyme. Seven altered proteins have been reported for the first time in the present work; coffilin-1 (Ctx), TI-225 (Ctx), peptidylprolyl isomerase A (Ctx), unnamed protein product (Ctx and Hippo), Syn2 protein (Hippo), and cysteine rich protein (crip2) (Hippo). Considering the obtained proteins, the time-series alteration of the most altered proteins in Ctx and Hippo was consistent with that of each ribosomal protein, while the expression of Syn2 protein (Hippo) increased with age. And there were fewer altered proteins in BS than in other regions, suggesting that the number of neuron functional units in the brain is rarely altered with aging BS. Consequently, the use of the FD-LC-MS/MS method enables the differential proteomics analysis of aging in segmented brain tissue sample and identification of 28 age-related altered proteins in mouse brain regions, seven of which have been reported for the first time in this work.

1.3.4 Differential Analysis of Skeletal Muscle Proteins in Fast Thoroughbred Horses

Since horses have extensive interactions with humans in both a wide variety of sports (e.g., horse racing) and in working for transport. In particular, the Thoroughbred horse is a remarkable animal, with both speed and endurance during running. To train a faster-running horse, the breeding and the training of Thoroughbred horses have been examined for more than 300 years [33–39]. To extend the applicability of the FD-LC-MS/MS method, we applied it to Thoroughbred horse muscle [11]. Thoroughbred horse muscle samples were analyzed and compared between trained and detrained horses. To quantify the changed protein expressions during exercise, the protein expressions obtained from the tissue after training was compared with that from the same sample after detraining. As a result, a great number of proteins were obtained as the training-to detraining ratio (+/–) of the changed proteins. Considering the differences observed above, we speculated that there might be relationship between the changes in protein expressions during exercise in each horse and physical ability of each horse, such as the running speed. The oxygen uptake (VO_2) values are related to endurance and are often used as an indicator of fast running [34, 35]. Using the VO_2max values as indicators of running speed, the correlations between VO_2max (post-training) and +/– ratio were calculated. Consequently, 16 protein peaks involved in energy production, such as aerobic and anaerobic energy supply, exhibited a good correlation coefficient. Identified proteins have contained muscle creatine kinase (M-CK) and glycolytic enzymes. The +/– ratios of both M-CK and glycolytic enzymes significantly correlated with the VO_2max values. M-CK and glycolytic enzymes are related to anaerobic energy supply, while the VO_2max value indicates aerobic capacity in theory. However, the product of anaerobic glycolysis, pyruvate, is transported into mitochondria, and then the metabolized product enters the aerobic TCA cycle; thus, an increase of the anaerobic glycolitic pathway might result in an increase of the aerobic TCA cycle, which produces ATP efficiently. Considering this cooperative between the enzymes of the anaerobic pathway, including M-CK and glycolysis, and the proteins involved in the aerobic pathway, it was speculated that these anaerobic pathways effectively activated the aerobic pathway in faster running horse. In conclusion, horse muscle tissues could be applicable for the analysis of the protein expression using the FD-LC-MS/MS method. Moreover, the changes in protein expressions during exercise were related to running speed of horses, suggesting that the anaerobic pathway effectively activated the aerobic pathway in faster-running horses.

1.4 Conclusion

The FD-LC-MS/MS method that we have developed has two unique features in using a fluorogenic reagent DAABD-Cl to derivatize proteins and HPLC to separate the derivatized proteins. The reagent gives low background fluorescence affording the sensitive detection of derivatized proteins, while HPLC contributes the reproducible quantitation of proteins. Actually, its application data has proved the method to be highly sensitive and reproducible for the peak heights and retention times, indicating that the method is accurate.

In each application (mice liver, breast cancer cell lines, mice brain and Thoroughbred horse muscle), a differential proteomics analysis has been performed. According to hepatitis-infected mouse liver analysis, the present method suggested to be worthwhile for clinical proteomics analysis. In breast cancer cell analysis, the protein expression levels in eight different cell lines were compared simultaneously, which has given results that suggested a novel pathway correlated to invasive and metastatic breast cancer. The analysis of segmented mouse brain demonstrated that the method is applicable for a small organ and the age-related proteins in mouse brain regions were elucidated. In Thoroughbred horse skeletal muscle analysis, the enzymes involved in the aerobic pathway have been obtained as altered proteins during exercises, suggesting that the anaerobic pathway effectively activated the aerobic pathway in faster-running horses.

In the future, it should be extended to differential proteomics analyses of many biological samples revealing the roles of certain proteins in the special biochemical events within living animals.

References

1. Benson, J. R., and Hare, P. E. (1975) O-phthalaldehyde: fluorogenic detection of primary amines in the picomole range. Comparison with fluorescamine and ninhydrin. *Proc. Natl. Acad. Sci. U. S. A.*, **72**, pp. 619–622.
2. Udenfriend, S., Stein, S., Bohlen, P., Dairman, W., Leimgruber, W., and Weigele, M. (1972) Fluorescamine: a reagent for assay of amino acids, peptides, proteins, and primary amines in the picomole range. *Science*, **178**, pp. 871–872.
3. Toriumi, C., and Imai, K. (2003) An identification method for altered proteins in tissues utilizing fluorescence derivatization, liquid chromatography, tandem mass spectrometry, and a database-searching algorithm. *Anal. Chem.*, **75**, pp. 3725–3730.
4. Masuda, M., Toriumi, C., Santa, T., and Imai, K. (2004) Fluorogenic derivatization reagents suitable for isolation and identification of cysteine-containing proteins utilizing high-performance liquid chromatography-tandem mass spectrometry. *Anal. Chem.*, **76**, pp. 728–735.
5. Masuda, M., Saimaru, H., Takamura, N., and Imai, K. (2005) An improved method for proteomics studies in C. elegans by fluorogenic derivatization, HPLC isolation, enzymatic digestion and liquid chromatography-tandem mass spectrometric identification. *Biomed. Chromatogr.*, **19**, pp. 556–560.
6. Asamoto, H., Ichibangase, T., Saimaru, H., Uchikura, K., and Imai, K. (2007) Existence of low-molecular-weight thiols in *Caenorhabditis elegans* demonstrated by HPLC-fluorescene detection utilizing 7-chloro-N-[2-(dimethylamino)ethyl]-2,1,3-benzoxadiazole-4-sulfonamide. *Biomed. Chromatogr.*, **21**, pp. 999–1004.
7. Ichibangase, T., Moriya, K., Koike, K., and Imai, K. (2007) A proteomics method revealing disease-related proteins in livers of hepatitis-infected mouse model. *J. Proteome Res.*, **6**, pp. 2841–2849.
8. Imai, K., Koshiyama A., and Nakata K. (2011) Towards clinical proteomics analysis, *Biomed. Chromatogr.*, **25**, pp. 59–64.
9. Imai, K., Ichibangase, T., Saitoh, R., and Hoshikawa, Y. (2008) A proteomics study on human breast cancer cell lines by fluorogenic derivatization-liquid chromatography/tandem mass spectrometry. *Biomed. Chromatogr.*, **22**, pp. 1304–1314.
10. Asamoto, H., Ichibangase, T., Uchikura, K., and Imai, K. (2008) Application of an improved proteomics method, fluorogenic derivatization-liquid

chromatography-tandem mass spectrometry, to differential analysis of proteins in small regions of mouse brain. *J. Chromatogr. A*, **1208,** pp. 147–155.

11. Ichibangase, T., and Imai, K. (2009) Application of fluorogenic derivatization-liquid chromatography-tandem mass spectrometric proteome method to skeletal muscle proteins in fast Thoroughbred horses. *J. Proteome Res.*, **8**, pp. 2129–2134.
12. Ichibangase, T., Moriya, K., Koike, K., and Imai, K. (2009) Limitation of immunoaffinity column for the removal of abundant proteins from plasma in quantitative plasma proteomics. *Biomed. Chromatogr.*, **23**, pp. 480–487.
13. Ichibangase, T., Saimaru, H., Takamura, N., Kuwahara, T., Koyama, A., Iwatsubo, T., and Imai, K. (2008) Proteomics of Caenorhabditis elegans over-expressing human alpha-synuclein analyzed by fluorogenic derivatization-liquid chromatography/tandem mass spectrometry: identification of actin and several ribosomal proteins as negative markers at early Parkinson's disease stages. *Biomed. Chromatogr.*, **22**, pp. 232–234.
14. Moriya, K., Yotsuyanagi, H., Shintani, Y., Fujie, H., Ishibashi, K., Matsuura, Y., Miyamura, T., and Koike, K. (1997) Hepatitis C virus core protein induces hepatic steatosis in transgenic mice. *J. Gen. Virol.*, **78** (Pt 7), pp. 1527–1531.
15. Moriya, K., Fujie, H., Shintani, Y., Yotsuyanagi, H., Tsutsumi, T., Ishibashi, K., Matsuura, Y., Kimura, S., Miyamura, T., and Koike, K. (1998) The core protein of hepatitis C virus induces hepatocellular carcinoma in transgenic mice. *Nat. Med.*, **4**, pp. 1065–1067.
16. Moriya, K., Nakagawa, K., Santa, T., Shintani, Y., Fujie, H., Miyoshi, H., Tsutsumi, T., Miyazawa, T., Ishibashi, K., Horie, T., Imai, K., Todoroki, T., Kimura, S., and Koike, K. (2001) Oxidative stress in the absence of inflammation in a mouse model for hepatitis C virus-associated hepatocarcinogenesis. *Cancer Res.*, **61**, pp. 4365–4370.
17. Moriya, K., Todoroki, T., Tsutsumi, T., Fujie, H., Shintani, Y., Miyoshi, H., Ishibashi, K., Takayama, T., Makuuchi, M., Watanabe, K., Miyamura, T., Kimura, S., and Koike, K. (2001) Increase in the concentration of carbon 18 monounsaturated fatty acids in the liver with hepatitis C. Analysis in transgenic mice and humans. *Biochem. Biophys. Res. Commun.*, **281**, pp. 1207–1212.
18. Dragani, T. A., Manenti, G., Sacchi, M. R., Colombo, B. M., and Della Porta, G. (1989) Major urinary protein as a negative tumor marker in mouse hepatocarcinogenesis. *Mol. Carcinog.*, **2**, pp. 355–360.

19. Ejiri, S. (2002) Moonlighting functions of polypeptide elongation factor 1: From actin bundling to zinc finger protein R1-associated nuclear localization. *Biosci. Biotechnol. Biochem.*, **66**, pp. 1–21.

20. Shiraishi, T., Mori, M., Tanaka, S., Sugimachi, K., and Akiyoshi, T. (1998) Identification of cystatin B in human esophageal carcinoma, using differential displays in which the gene expression is related to lymph-node metastasis. *Int. J. Cancer*, **79**, pp. 175–178.

21. Wu, S. J., and Li, S. E. (1988) Studies on the mechanism of carcinogenesis (IX)--Differential expression of proliferating and tissue-specific enzyme genes during hepatocarcinogenesis. *Sci. Sin. B*, **31**, pp. 590–595.

22. Gruvberger-Saal, S. K., Cunliffe, H. E., Carr, K. M., and Hedenfalk, I. A. (2006) Microarrays in breast cancer research and clinical practice — the future lies ahead. *Endocr.-Relat. Cancer*, **13**, pp. 1017–1031.

23. Abba, M. C., Sun, H., Hawkins, K. A., Drake, J. A., Hu, Y., Nunez, M. I., Gaddis, S., Shi, T., Horvath, S., Sahin, A., and Aldaz, C. M. (2007) Breast cancer molecular signatures as determined by SAGE: correlation with lymph node status. *Mol. Cancer Res.*, **5**, pp. 881–890.

24. Hondermarck, H., Vercoutter-Edouart, A. S., Revillion, F., Lemoine, J., El-Yazidi-Belkoura, I., Nurcombe, V., and Peyrat, J. P. (2001) Proteomics of breast cancer for marker discovery and signal pathway profiling. *Proteomics*, **1**, pp. 1216–1232.

25. Osborne, C. K. (1998) Tamoxifen in the treatment of breast cancer. *N. Engl. J. Med.*, **339**, pp. 1609–1618.

26. Colomer, R., Shamon, L. A., Tsai, M. S., and Lupu, R. (2001) Herceptin: From the bench to the clinic. *Cancer Invest.*, **19**, pp. 49–56.

27. Stuhler, K., Pfeiffer, K., Joppich, C., Stephan, C., Jung, K., Muller, M., Schmidt, O., van Hall, A., Hamacher, M., Urfer, W., Meyer, H. E., and Marcus, K. (2006) Pilot study of the Human Proteome Organisation Brain Proteome Project: Applying different 2-DE techniques to monitor proteomic changes during murine brain development. *Proteomics*, **6**, pp. 4899–4913.

28. Focking, M., Boersema, P. J., O'Donoghue, N., Lubec, G., Pennington, S. R., Cotter, D. R., and Dunn, M. J. (2006) 2-D DIGE as a quantitative tool for investigating the HUPO Brain Proteome Project mouse series. *Proteomics*, **6**, pp. 4914–4931.

29. Seefeldt, I., Nebrich, G., Romer, I., Mao, L., and Klose, J. (2006) Evaluation of 2-DE protein patterns from pre- and postnatal stages of the mouse brain. *Proteomics*, **6**, pp. 4932–4939.

30. Carrette, O., Burkhard, P. R., Hochstrasser, D. F., and Sanchez, J. C. (2006) Age-related proteome analysis of the mouse brain: a 2-DE study. *Proteomics*, **6**, pp. 4940–4949.

31. Frohlich, T., Helmstetter, D., Zobawa, M., Crecelius, A. C., Arzberger, T., Kretzschmar, H. A., and Arnold, G. J. (2006) Analysis of the HUPO Brain Proteome reference samples using 2D DIGE and 2D LC-MS/MS. *Proteomics*, **6**, pp. 4950–4966.

32. Tsugita, A., Kawakami, T., Uchida, T., Sakai, T., Kamo, M., Matsui, T., Watanabe, Y., Morimasa, T., Hosokawa, K., and Toda, T. (2000) Proteome analysis of mouse brain: Two-dimensional electrophoresis profiles of tissue proteins during the course of aging. *Electrophoresis*, **21**, pp. 1853–1871.

33. Yamano, S., Eto, D., Hiraga, A., and Miyata, H. (2006) Recruitment pattern of muscle fibre type during high intensity exercise (60–100% VO_2max) in Thoroughbred horses. *Res. Vet. Sci.*, **80**, pp. 109–115.

34. Eto, D., Yamano, S., Mukai, K., Sugiura, T., Nasu, T., Tokuriki, M., and Miyata, H. (2004) Effect of high intensity training on anaerobic capacity of middle gluteal muscle in Thoroughbred horses. *Res. Vet. Sci.*, **76**, pp. 139–144.

35. Eaton, M. D., Hodgson, D. R., Evans, D. L., and Rose, R. J. (1999) Effects of low- and moderate-intensity training on metabolic responses to exercise in Thoroughbreds. *Equine. Vet. J. Suppl.*, **30**, pp. 521–527.

36. Harkins, J. D., Beadle, R. E., and Kamerling, S. G. (1993) The correlation of running ability and physiological variables in Thoroughbred racehorses. *Equine. Vet. J.*, **25**, pp. 53–60.

37. Yamano, S., Eto, D., Sugiura, T., Kai, M., Hiraga, A., Tokuriki, M., and Miyata, H. (2002) Effect of growth and training on muscle adaptation in Thoroughbred horses. *Am. J. Vet. Res.*, **63**, pp. 1408–1412.

38. Roneus, N., and Essen-Gustavsson, B. (1997) Skeletal muscle characteristics and metabolic response to exercise in young standardbreds. *Am. J. Vet. Res.*, **58**, pp. 167–170.

39. Poso, A. R. (2002) Monocarboxylate transporters and lactate metabolism in equine athletes: A review. *Acta Vet. Scand.*, **43**, pp. 63–74.

Chapter 2

Two-Dimensional Difference Gel Electrophoresis

Viola Ruddat
GE Healthcare, Piscataway, NJ 08855, USA
viola.ruddat@ge.com

Finding differentially expressed proteins is an important component in many areas of biological research. A common approach is to separate the proteins by chromatography or electrophoresis followed by identification using mass spectrometry. Two-dimensional electrophoresis is commonly employed as the electrophoretic method and enables very-high-resolution separations. In the mid 1990s, the term proteomics was coined to describe this research area, establishing an additional expression-based dimension to genomics.

2.1 Introduction

Since its inception in 1975, 2D electrophoresis has been one of the most commonly used techniques for differential expression

Quantitative Proteome Analysis: Methods and Applications
Edited by Kazuhiro Imai and Sam F. Y. Li

ISBN 978-981-4316-51-4 (Hardcover), 978-981-4316-98-9 (eBook)
www.panstanford.com

proteomics. This technique, while often very successful, comes with certain inherent caveats. It tends to require a significant amount of hands on time, and, for quantitation purposes, technical replicates are usually required in addition to biological replicates because of variations introduced by the technique itself. Because of that, many researchers today forgo 2D electrophoresis and turn to liquid chromatography-mass spectrometry (LC-MS) instead. However, LC-MS and 2D electrophoresis have very different advantages, and despite many efforts, there is still a significant amount of information to be gained by 2D electrophoresis that cannot be obtained as easily by LC-MS. One of these advantages is the ease of visualization of proteins with the same molecular weight but different isoelectric points (pIs), allowing for the simple location of proteins with post-translational modifications (PTMs). Also, when carried out correctly, the quantitative comparison of many biological replicates is greatly simplified.

2.2 Two-Dimensional Difference Gel Electrophoresis

Many improvements have been made to the traditional 2D gel technique to enhance the automation and throughput. One of the greatest advances was the introduction, in the early 1990s, of commercially available immobilized pH gradients on strips instead of the classical carrier ampholyte based tube gels. This approach provided both enhanced reproducibility and ease of handling over the tube gels of the past. Other product advances include the ability to multiple strips and gels in parallel, pre- made buffers, and better software, all of which have helped significantly reduce the hands-on time. Reproducibility, however, remained an issue. The introduction of fluorescent dyes, that can be used to label the lysine or cysteine residues of proteins prior to electrophoresis and allows for multiplexed 2D helps overcome this issue. The charge- and molecular weight–matched fluorescent dyes allow for the same protein from different samples (labeled with different dyes) to co-migrate to the same position in the gel. This technology is called two-dimensional difference gel electrophoresis (2D DIGE). 2D DIGE introduced the ability to remove non-biological variation from the system, by including an internal standard that allows for global normalization[1] and is at the core of this technology.[2] The internal standard is a pooled mixture of all the samples in a given experiment, and an aliquot of this same mixture is applied to each gel in the experiment.[3] This internal standard is commonly labeled with the Cy2 dye. Since one of the samples is now in effect the same on all gels, it can be used to normalize out experimental variation between gels. The internal standard also allows for higher quality automated matching, since in effect, every gel contains the same sample mixture in the Cy2 channel.

Additional advantages of 2D DIGE are a significant reduction of the number of gels that need to be run and increased limits of detection and dynamic range due to the fluorescent Cy dyes used.

2.2.1 Minimal Dyes

In the minimal labeling approach, three charge- and molecular weight–matched dyes are available, which excite at 488 nm (Cy™2), 532 nm (Cy3), and 635 nm (Cy5) respectively. The term minimal refers to the strategy to use stoichiometry to ensure no protein is ever labeled more than once. In fact, only 1–3% of lysine residues (on any given protein species) are labeled. Due to the separation in excitation and emission spectra, it is possible to visualize all three dyes without cross talk using an appropriate imaging system. In this approach, every gel contains two real samples (Cy3 and Cy5), for example, a treated and a control sample, and the internal standard in the third channel, usually Cy2.

2.2.2 Saturation Dyes

An alternative approach to using the minimal labeling concept is to use specially designed saturation dyes that are matched for both charge and molecular weight. These dyes conjugate to reduced cysteine residues on the proteins and the strategy employed is usually to optimize the reduction of these amino acids to allow maximal binding to all available cysteine groups. This results in an enhanced limit of detection allowing for a reduced amount of sample being necessary for labeling (10 μg rather than the 50 μg used for minimal labeling). Many applications make use of this approach; one of the main applications is where sample is limited, such as with laser capture microdissection.[4,5]

2.3 Quantitative Aspects of 2D DIGE

In the early years of 2D electrophoresis, statistics were unfortunately not usually a consideration. This has led to unreliable and poorly reproducible results. Much effort has been applied by the proteomics community to investigate the need for statistics, and trying to establish guidelines, one such organization is the Association of Biomolecular Resource Facilities (ABRF).

All quantitative measurements are usually based on relative quantitation. Relative quantitation is based on the assumption that every sample contains the same amount of total protein. Most 2D software analysis packages will include a tool that allows the cross normalization of loading differences. In the case of DeCyder™, this is accomplished in a module called DIA (Difference-In Gel Analysis), which takes advantage of a normal distribution to normalize loading differences. For special cases, mentioned under applications in a later section, this normalization procedure can also be accomplished by spiking in known amounts of specific proteins.

The most basic requirement for statistical calculations is the use of replicates, which is now thankfully becoming more common for 2D users. Originally, both technical replicates (to handle experimental variation) and biological replicates (to handle inherent biological variation) were needed. While 2D DIGE removes the need for technical replicates, biological replicates are still needed to remove biological variation. When replicates are used, the *p*-value cut offs selected often ignore the problem of multiple comparisons. In a typical gel, about 1000 real proteins can be detected. If we are using a t-test, we are in fact applying 1000 univariate tests. In the past, a significance level of $p = 0.05$ was often mistakenly used. Without adjustments, that means that the probability of type I errors (false positives) is 5% of 1000, or 50 false positives. In order to control the overall *p*-value, an adjustment to the *p*-value is required[6] to reflect the same probability applied to 1000 spots.

Type I errors are not the only consideration; type II errors (false negatives) have to be controlled as well. Depending on the desired minimum abundance differences to be detected, false negatives have to be controlled by replicates.[7]

The number of replicates, especially for clinical samples, quickly becomes overwhelming, and pooling strategies can be applied that allow a reduction in the number of gels, while still being able to exclude some outliers.[8] An excellent overall review of 2D DIGE with respect to quantitation was written by Lilley and Friedman.[9] Another aspect of 2D DIGE is the potential of labeling bias, which can be avoided in one of two ways, either by performing a dye flip/reverse labeling, or by using a two-dye approach, the latter having the additional advantage of resulting in a decrease in data scatter (reduction of coefficient of variance).[10]

The use of advanced statistics, particularly the use of multivariate statistics, has been commonly employed for microarray studies, and recently has also been used in the proteomics community. Researchers use statistical tools integrated into analysis software such as DeCyder EDA[11] as well as external biostatistical tools to apply to proteomics data.[12] This approach allows for easier analysis of more than two sample types, unknown groupings can be discovered, outliers eliminated, and data visualized.

Generally, for good quantitation, appropriate software is critical. Requirements are stable and reliable detection and matching algorithms, the capability to co-detect, and the ability to take advantage of the internal standard for normalization. Software packages that are able to carry out statistical calculations for the user will enhance robustness of results.

2.4 Application Examples

Differential proteomics utilizing 2D DIGE in particular are used in a large variety of applications. Much research is performed with regard to finding biomarkers for cancers, prion disease, schizophrenia,[13] heart disease,[14] and Alzheimer's disease,[15] to name a few. The sample type used extends from cell lines to tissues and serum derived from any number of model animals to clinical samples. Biomarkers can be used for diagnostic purposes to elucidate pathways of disease progression and to identify potential drug targets. In preclinical studies, the impact on the proteome by the addition of drugs to cell lines and animals, can also be monitored with 2D DIGE. The particular usefulness of 2D DIGE over other differential proteomics techniques is its capability to find abundance differences down to 20%. Also, with the inherent biological variation (especially with clinical samples), the comparison of many replicates is a necessity.

Other research areas include stem cell research, agricultural research, and quality control in the production of biopharmaceuticals. The following section will discuss some of these topics in more detail.

2.4.1 Cancer-Related Research

One of the most common areas of expression-based proteomics is located in cancer research. One in every three people is estimated to be diagnosed with some kind of cancer in their lifetime. Proteomics can help to build a foundation to better understand cancer, and use these findings to develop and improve diagnosis and treatment. There are three main areas: finding biomarkers, identifying potential drug targets, and evaluating the impact of drugs on the proteome. Finding biomarkers is especially crucial to both determine subclasses of cancer to allow for personalized treatment and for early recognition.

Cancer biomarkers and disease progression investigated with the aid of 2D DIGE exceeds well over 100 publications based on highwire: http://highwire.standard.edu. Cancers researched include lung cancer,[16] colorectal cancer,[17] breast cancer,[18] and prostate cancer, to name some of the most deadly ones. Prostate cancer is a typical example to illustrate the need and usefulness of differential proteomics. Prostate cancer is the leading cause of cancer-related deaths in America. Gleason grading is currently the primary method for prediction, with only a few protein-based biomarkers available, prostate specific antigen (PSA) being the most common. Unfortunately, PSA results lead to both false negatives (one in three prostate cancer cases do not result in higher PSA levels) and false positives leading to unnecessary invasive procedures.[19] More biomarkers relating to cancer progression, early detection, and differentiating of subclasses are desirable. While no biomarkers or panel of biomarkers for reliable early detection of prostate cancer are currently available, the use of 2D DIGE, particularly with the use of appropriate statistics, has resulted in good progress. Fun *et al.* applies 2D DIGE data in conjunction with nuclear magnetic resonance (NMR) to look at the metabolic profile in serum of prostate cancer patients with Gleason scores 5–7.[20] Other research establishes common pathways of disease progression by combining 2D DIGE data with systems biology,[21] implicating proteins involved in calcium regulation,[22] and confirmation of early stage biomarkers established by orthogonal methods.[23] Cancer research is the single most common research topic for 2D DIGE.

2.4.2 Other Diseases

Cancer is not the only disease that requires molecular biomarkers, as symptoms can be non-specific (Alzheimer's disease), or can only be established at the death of a patient (Creutzfeldt–Jakob disease [CJD]).[24] For other diseases, differential proteomic studies during their progression can help elucidate a new view of causes and hence treatment (addiction,[25] schizophrenia.[26]) As with cancer, these studies require many clinical samples to be compared to one another, and 2D DIGE is currently the most quantitative way to do this.

2.4.3 Agricultural Research

Proteomics in agricultural research is also gaining great importance. An area of great concern has been mad cow disease, a prion disease. Currently, this disease can only be diagnosed after butchering, which allows for possible contamination of adjacent meat. Ideally, it should be possible to test easily accessible fluid such as urine for diagnostic markers ahead of time, and such studies are currently being investigated.[27] Another topic of increasing importance is genetically modified foods. Food safety and the unintended side effects of genetic modifications of both crops[28] and livestock[29] can be partially elucidated by identifying the change in protein abundance combined with systems biology.

Another great area of interest for differential proteomics research is the investigation of crop plants, especially related to stress-induced response.[30]

2.4.4 DIGE and RNA

For the past 10 years, efforts to correlate mRNA and protein expression have been abundant. Depending on the proteome investigated, transcription and translation correlates at best at about 40%. Some reasons for this are experimental difficulties, especially regarding the accurate quantitation of protein levels and experimental challenges of protein identification. Biological causes play a larger role.

One of the main biological reasons is that translation is not only dependent on mRNA concentration but also on rate of translation.[31]

Other discrepancies are based on secondary expression events. A notable study by Muniyappa *et al.* illustrates the use of 2D DIGE.[32] They found an overexpression of miRNA-29a to have protective effects. Comparing the predicted changes in the proteome, only correlated 20% of the time; however, this could be due to biological reasons such as most proteins identified are secondary targets, or reduced importance of the canonical region. Regardless, this approach, where a vector is added to increase one particular RNA sequence, allows researchers to better understand the cause and effect of transcription and translation.

2.4.5 Drug Studies

The impact of drugs on the proteome can be monitored with the aid of differential proteomics. The levels of protein expression can be monitored for their beneficial or deleterious effects.

An example comparing the proteome of patients responding to a certain drug with those who are resistant is shown in a clinical study.[33] Other studies are often performed in cell lines or animal models, and thus the number of replicates required is less than that for clinical studies, but many different sample types often need to be compared.[34] Often, both dosage and time course are studied in parallel. Both *in vivo* and *in vitro* studies offer important answers regarding drug resistance, side effects, and insight into the biological pathways affected by certain drugs.

2.4.6 Stem Cell Research

Proteomic studies are useful in many areas of stem cell research. Typical areas of stem cell research relate to cardiovascular and neurological repair, treatment of arthritis, spinal cord injuries, and Alzheimer's disease. Some of the challenges in the use of stem cells are their origin (embryonic stem cells, adult stem cells, etc.) and the ability to verify successful differentiation without undifferentiated cell impurities. Some markers, often membrane proteins, have been established to date.[35] Differential proteomics, both 2D DIGE and LCMS can be used to find panels of markers to determine stem cell differentiation. To investigate the possibility of using stem cell lines that are not embryonic, it needs to be verified that the proteome of cells differentiated from different types of stem cells or stem cell lines are in fact the same.[36] 2D DIGE can also be used to find markers in cancer cells that lead to the differentiation of cancer stem cells, or identify the molecular mechanism that leads to proliferation and differentiation of cancer stem cells.[37] In addition, proteomic studies also offer a way to address questions regarding stem cell response to disease/injury. There are many other uses of proteomics in stem cell research and is an expanding field.

2.4.7 Biopharmaceutical: Quality Control and Process Development

The examples discussed so far have been relatively common in studying differential expression. However, 2D DIGE also has a use in the biopharmaceutical industry.

The past decade has seen a dramatic increase in biopharmaceutical research with an average of 15 biopharmaceuticals approved by the FDA annually. These drugs range from monoclonal antibodies and vaccines to interleukins and interferons and have been used successfully to treat numerous diseases, including cancer, autoimmune disease, inflammation, asthma, colitis, and others (http://www.biopharma.com/approvals_2010.html). Two recent examples are infliximab for arthritis and trastuzumab for breast cancer.

The development of biopharmaceuticals undergoes multiple stages. First drug targets are identified and best "binders" are optimized. In the next stage, the preliminary drug goes through clinical trials and formulation studies. The last stage is the development and scale-up of production methods. Biopharmaceutical manufacturing requires the selection of an appropriate host species, such as Chinese Hamster Ovary (CHO) cells or a bacterial host, such as *Escherichia coli*, a decision usually dictated by the best yield of the biomolecule of interest. The final stage requires the optimization of growth, harvest, and purification conditions.

The purification of biopharmaceuticals, whether from CHO cells or bacteria, presents a particular challenge. For most research science purposes, purity requirements are rarely greater than 95%.

For biopharmaceuticals, the regulatory requirements are strict, with generally less than 1% impurities allowed by the regulatory authorities, like the FDA. Actual levels depend on the particular host cell, the drug itself, as well as its dosage and application. For example, drugs administered over a long period of time or at high dosages have extremely high purity requirements. Certain animal derived media additives for cell growth pose a potential additional risk. Consistent purity guidelines are elusive, since limited adverse effects of residual host cell protein or animal additives have been reported. However, a cautionary attitude prevails due to wise concerns of possible allergic reactions and other toxic effects.[38]

At this time, neither has a general strategy specific to the removal of host cell protein (HCP) and media additives been developed nor are general methods for purity checking approved.

Purity is usually checked with an enzyme-linked immunosorbent assay (ELISA), specifically developed for every process, and every drug. However, ELISA will only detect overall residual HCP, and the same signal could represent a large range of HCP proteins. Hence, orthogonal methods, such as sodium dodecyl sulfate polyacrylamide gel electrophoresis (SDS PAGE) and 2D electrophoresis are often used for better resolution. While SDS PAGE is still used more commonly, it has shortcomings, in missing the resolution afforded by PI and often the use of less sensitive post stains (e.g., Coomassie). As impurities (as discussed above) are of great concern, other techniques are being considered more and more.

Among such techniques are 2D in general and 2D DIGE in particular. In a proof-of-principle experiment, monoclonal antibodies (MAb) were expressed in CHO cells in different cell culture media followed by affinity purification. Harvest, flow-through, and eluate were tested for residual HCP with ELISA and 2D DIGE. The results showed that while overall HCP levels were determined to increase with ELISA and most spots were shown to increase with 2D DIGE based on the different growing conditions, individual proteins could also be shown to decrease with 2D DIGE, a trend that would have been lost with ELISA alone.[39]

In addition to optimizing media and rationally designing purification, the 2D DIGE approach can be used to determine optimal cell lines, viability, temperature and aggregation in downstream process development.[40]

2.5 Special Applications

2.5.1 Pre-Fractionation

A common issue in proteomics is resolving power. While 2D electrophoresis is superior to other methods in this respect, proteins can still co-migrate in the gel. Different methods of pre-fractionation have been employed in the past to help rectify this problem. Examples of pre-fractionation techniques include various chromatography techniques (ion exchange, reverse phase, hydrophobic interaction chromatography, HIC), preparative isoelectric focusing (IEF), depletion of particular abundant proteins such as albumin, enrichment of glycosylated or phosphorylated proteins,[41] and the fractionation of cellular organelles.[42] While this challenge is common to all 2D electrophoresis techniques, 2D DIGE provides a particularly straightforward way to verify reproducibility as illustrated in the next sections.

A caveat of every sample preparation step, especially pre-fractionation, is the possible introduction of additional experimental variation. Using 2D DIGE is a great way to validate that pre-fractionation techniques do not add additional variation. As an example, we have pre-fractionated an *E. coli* sample labeled with Cy2, Cy3, and Cy5 dyes using anion exchange chromatography.[43] We carried out a same/same experiment with three replicate runs and quantitatively analyzed the resulting three replicates of 24 chromatography fractions using 2D DIGE and the associated analysis software. With this procedure, we examined nine pooled fractions and compared them with an un-fractionated sample. The total number of spots detected was 4197 from all fractions combined, versus 820 spots from the un-fractionated sample. We were able to show that the same protein, whether unlabeled or labeled with different CyDye DIGE fluors, eluted from the anion exchange column in the same fraction. In addition, we demonstrated that the pre-fractionation using anion exchange chromatography is highly reproducible under these conditions, and that this is thus a valid method for pre-fractionation of complex protein.

2.5.2 DIGE and Sample Preparation

Another example of using 2D DIGE technology for determining the best sample preparation method is by comparing different extraction buffers (Data file 28-9488–39 AA, GE Healthcare). The 2D DIGE multiplexing technique was used to estimate which sample extraction buffers would result in the extraction of the most proteins from various tissues.

2.5.3 Cell Surface Labeling with 2D-DIGE

Traditional cellular fractionation carries out actual physical separation of different organelles. 2D DIGE is able to visualize proteins labeled on the cell surface and compare it with those proteins inside the cell.[44]

This technique takes advantage of the fact that for labeling to occur the pH has to be higher than 8. So, when labeling intact cells, if there is any chance that a label can enter the cell, they will not label the proteins on the inside, since the biological pH is closer to 7. If another fraction of the same cells is first lysed, the pH adjusted, and then labeled (with a different CyDye), and both fractions are run on the same gel, intra cellular proteins and surface proteins can be compared.[45]

2.5.4 Membrane Proteins

One of the issues associated with 2D gels (particularly immobiline-based 2D work) is the variability in success of recovery of the very hydrophobic proteins. One of the ways to work around this problem is to work with blue native gel electrophoresis coupled with SDS PAGE–this allows for the ability to look at protein complexes and can be further utilized by using the 2D DIGE technology.[46]

Another option to blue native for the first dimension separation prior to SDS PAGE is to use cationic detergent such as 16 BAC (benzyldimethyl-n-hexadecylammoniumchloride) to help with solubilizing the membrane proteins.[47]

2.5.5 Study of Redox Potential

Another approach using the saturation dyes is to study redox potentials. In the paper by Hurd *et al.*,[48] they studied reactive oxygen species in the mitochondrial respiratory chain and monitored how their levels affected the thiol redox state of mitochondrial proteins using these saturation dyes.

2.6 Conclusion

2D DIGE greatly expands on traditional 2D. Most importantly, the pre-labeling technique allows for multiplexing. This enables much improved quantitation leading to reproducible results. In addition, this pre-labeling approach gives rise to new application areas not available to traditional 2D.

References

1. Alban A., David S.O., Bjorkesten L., Andersson C., Sloge E., Lewis S., Currie I. (2003) A novel experimental design for comparative two-dimensional gel analysis: two-dimensional difference gel electrophoresis incorporating a pooled internal standard. *Proteomics*, **3**, 36–44.
2. Marouga R., David S., Hawkins E. (2005) The development of the DIGE system: 2D fluorescence difference gel analysis technology. *Anal. Bioanal. Chem.*, **382**, 669–678.
3. Beckett P. (2012) The basics of 2D DIGE. In: Cramer R., Westermeier R. Methods in Molecular Biology Volume 854, Difference Gel Electrophoresis (DIGE). Humana Press, Springer. p9–16.
4. Kondo T. (2008) Tissue proteomics for cancer biomarker development: laser microdissection and 2D-DIGE. *BMB Rep.*, **30**, 626–634.
5. Kondo T., Hirohashi S. (2006) Application of highly sensitive fluorescent dyes (CyDye DIGE Fluor saturation dyes) to laser microdissection and two-dimensional difference gel electrophoresis (2D-DIGE) for cancer proteomics. *Nat.* Protocols, 1, **6**, 2940–2956.
6. Karp N. A., McCormick P. S., Russell M. R., Lilley K. S. (2007) Experimental and statistical considerations of avoid false conclusions in proteomics studies using differential in-gel electrophoresis. *Mol. Cell. Proteomics*, **6**, 1354–1364.
7. Karp N. A., Lilley K. S. (2007) Design and analysis issues in quantitative proteomics studies. *Proteomics*, **7**, 42–50.
8. Karp N. A., Lilley K. S. (2009) Investigating sample pooling strategies for DIGE experiments to address biological variability. *Proteomics*, **9**, 388–397.
9. *Lilley* K. S., *Friedman* D. B. (2004) All about DIGE: quantification technology for differential-display 2D-gel proteomics. *Expert Rev. Proteomics*, **4**, 401–409.

10. Kreil D. P., Karp N. A., Lilley K. S. (2004) DNA microarray normalization methods can remove bias from differential protein expression analysis of 2D difference gel electrophoresis results. *Bioinformatics*, **20**, 2026–2034.

11. Friedman D. B., Lilley K. S. (2008) Optimizing the difference gel electrophoresis (DIGE) technology. *Methods Mol. Biol.*, **428**, 93–124.

12. Anderson T. J., Tchernyshoyov I., Diez R., Cole R. N., Geman D., Dang C. V., Winslow R. L. (2007) Discovering robust protein biomarkers for disease from relative expression reversals in 2-D DIGE data. *Proteomics*, **7**, 1197–1207.

13. Prabakaran S., Wengenroth M., Lockstone H. E., Lilley K., Leweke F. M., Bahn S. J. (2007) 2-D DIGE analysis of liver and red blood cells provides further evidence for oxidative stress in schizophrenia. *J Proteome Res.*, **6**(1), 141–149.

14. Chugh S., Suen C., Gramolini A. (2010) Proteomics and mass spectrometry: What have we learned about the heart? *Curr. Cardiol. Rev.*, **6**, 124–133.

15. Maarouf C. L., Andacht T. M., Kokjohn T. A., Castaño E. M., Sue L. I., Beach T. G., Roher A. E. (2009) Proteomic analysis of Alzheimer's disease cerebrospinal fluid from neuropathologically diagnosed subjects. *Curr. Alzheimer Res.*, **6**, 399–406.

16. Murphy L., Clynes, M., Keenan J. (2007) Proteomic analysis to dissect mitoxantrone resistance-associated proteins in a squamous lung carcinoma. *Anticancer Res.*, **27**, 1277–1284.

17. Barderas, R., Babel, I., Casal J. I., (2010) Colorectal cancer proteomics molecular characterization and biomarker discovery. *Proteomic. Clin. Appl.*, **4**, 159–178.

18. Ambrosino C., Tarallo R., Bamundo A., Cuomo D., Franci G., Nassa G., Paris O., Ravo M., Giovane A., Zambrano N., Lepikhova T., Jänne D. A., Baumann M., Nyman T. A., Cicatiello L., Weisz A., (2010) Identification of a hormone-regulated dynamic nuclear actin network associated with estrogen receptor α in human breast cancer cell nuclei. *Mol. Cell. Proteomic.*, **9**, 1352–1367.

19. Schröder F. H. (2009) Review of diagnostic markers for prostate cancer. *Cancer Res.*, **181**, 173–182.

20. Fan, Y., Murphy, T. B., Byrne, J. C., Brennan, L., Fitzpatrick, J. M., Watson, R. W. G. (2011) Applying random forests to identify biomarker panels in serum 2D-DIGE data for the detection and staging of prostate cancer. *J. Proteomic. Res.*, **10**, 1361–1373.

21. Ummanni, R., Mundt F., Pospisil H., Venz S., Scharf C., Barett C., Fälth M., Köllermann J., Walther R., Schlomm T., Sauter G., Bokemeyer C., Sültmann H., Schuppert A., Brümmendorf T. H., Balabanov S. (2011) Identification of clinically relevant protein targets in prostate cancer with 2D-DIGE coupled mass spectrometry and systems biology network platform. *PLoS One,* **6**, e16833.

22. Ruddat V. C., Whitman S., Klein R. D., Fischer S. M., Holman T. R. (2005) Evidence for downregulation of calcium signaling proteins in advanced mouse adenocarcinoma. *Prostate,* **64**, 128–138.

23. Byrne J. C., Downes M. R., O'Donoghue N., O'Keane C., O'Neill A., Fan Y., Fitzpatrick J. M., Dunn M. Watson R. W. (2009) 2D-DIGE as a strategy to identify serum markers for the progression of prostate cancer. *J. Proteome Res.,* **8**, 942–57.

24. Brechlin P., Jahn O., Steinacker P., Cepek L., Kratzin H., Sehnert S., Jesse S., Mollenhauer B., Kretzschmar H. A., Wilfang J., Otto M., (2008) Cerebrospinal fluid-optimized two-dimensional difference gel electrophoresis (2-D DIGE) facilitates the differential diagnosis of Creutzfeldt–Jacob disease. *Proteomics,* **8**, 4357–4366.

25. Williams K., Wu T., Colangelo C., Nairn A. C., (2004), Recent advances in neuroproteomics and potential application to studies of drug addiction. *Neuropharmacology,* **47**, 148–166.

26. Prabakaran S., Wengenroth M., Lockstone H. E., Lilley K., Leweke F. M., Bahn S. (2007) 2-D DIGE Analysis of lever and red blood cells provides further evidence for oxidative stress in schizophrenia. *J. Proteome Res.,* **6**, 141–149.

27. Plews M., Lamoureux L., Simon S. L. R., Graham C., Ruddat V. C., Czub S., Knox J. D. (2011) Factors affecting the accuracy of urine-based biomarkers of BSE. *Proteome Sci.,* **9**, 6.

28. Cellini F., Chesson A., Colquhoun I., Constable A., Davies H. V., Engel K. H. , Gatehouse A. M. R., Kärenlampi S., Kok E. J., Leguay J., Lehesranta S., Noteborn H. P. J. M., Pedersen J., Smith M. (2004) Unintended effects and their detection in genetically modified crops. *Food Chem. Toxicol.,* **42**, 1089–1125.

29. Nakamura R., Nakamura R., Nakano M., Arisawa K., Ezaki R., Horiuchi H., Teshima R. (2010) Allergenicity study of EGFP-transgenic chicken meat by serological and 2D-DIGE analysis. *Food Chem. Toxicol.,* **48**, 1302–1310.

30. Yohei N., Nouri M. Z., Komatsu S. (2011) Quantitative proteomic analyses of crop seedlings subjected to stress conditions: a commentary. *Phytochemistry,* **72**, 1263–1272.

31. Vogel C., de Sousa Abreu R., Ko D., Le S. Y., Shapiro B. A., Burns S. C., Sandhu D., Boutz D. R., Marcotte E. M., Penalva L. O. (2010) Sequence signatures and mRNA concentration can explain two-thirds of protein abundance variation in a human cell line. *Mol. Syst. Biol.,* **6**, 400.

32. Muniyappa M. K., Dowling P., Henry M., Meleady P., Doolan P., Gammell P., Clynes M., Barron N. (2009) miRNA-29a regulates the expression of numerous proteins and reduces the invasiveness and proliferation of human carcinoma cell lines. *Eur. J. Cancer,* **45**, 3104–3118.

33. Rajpal R., Dowling P., Meiller J., Clarke C., Murphy W. G., O'Connor R., Kell M., Mitsiades C., Richardson P., Anderson K. C., Clynes M., O'Gorman P. (2011) A novel panel of protein biomarkers for predicting response to thalidomide-based therapy in newly diagnosed multiple myeloma patients. *Proteomics,* **11**, 1391–1402.

34. Liang X., Hajivandi M., Veach D., Wisniewski D., Clarkson B., Resh M. D., Pope R. M. (2006) Quantification of change in phosphorylation of BCR-ABL kinase and its substrates in response to Imatinib treatment in human chronic myelogenous leukemia cells. *Proteomics,* **6**, 4554–4564.

35. Intoh A., Kurisaki A., Yamanaka Y., Hirano H., Fukuda H., Sugino H., Asashima M. (2009) Proteomic analysis of membrane proteins expressed specifically in pluripotent murine embryonic stem cells. *Proteomics,* **9**, 126–137.

36. Stastna M., Abraham M. R., Van Eyk J. E., (2009). Cardiac stem/ progenitor cells, secreted proteins, and proteomics. *FEBS Lett.*, 583, 1800–1807.

37. Scatena R., Bottoni P., Giardina B. (2008) Modulation of cancer cell line differentiation: a neglected proteomic analysis with potential implications in pathophysiology, diagnosis. *Proteomic. Clin. Appl.,* **2**, 229–237.

38. Champion K., Madden H., Dougherty J., Shacter E. (2005) Defining your product profile and maintaining control over it, part 2. *BioProcess Tech.,* **3**, 52–57.

39. Hagner Å., McWhirter, S. Grimsby, A. Jorsback, T. Björkman, L. Kask, L. Björkesten, Malmquist G. (2010) A 2-D DIGE approach for characterization of host cell protein patterns improves process understanding. *Dis. Matter.,* **11**, 18–19.

40. Jin M., Szapiel N., Zhang J., Hickey J., Ghose S. (2010) Profiling of host cell proteins by two-dimensional difference gel electrophoresis (2D-DIGE): Implications for downstream process development, *Biotechnol. Bioeng.,* **105,** 306–316.

41. Stasyk T., Huber L. A. (2004) Zooming in: fractionation strategies in proteomics. *Proteomics*, **4**, 3704–3716.

42. Gauthier D., Lazure C. (2008) Complementary methods to assist subcellular fractionation in organellar proteomics. *Exp. Rev. Proteomic.*, **5**, 603–617.

43. Ruddat V. C., Beckett P., Rozanas C. R., Asbury G. R., (2007) P179-M A verification of novel experimental design for comparative two-dimensional gel analysis: pre-labeled samples fractionated on anion exchange column followed by 2D DIGE. *J. Biomol. Tech.*, **18**, 62.

44. Hagner-McWhirter A., Winkvist M., Bourin S., Marouga R. (2008) Selective labelling of cell-surface proteins using CyDye DIGE Fluor minimal dyes. J. *Vis. Exp.*, **21**, 945.

45. Mayrhofer C., Krieger S., Allmaier G., Kerjaschki D., (2006), DIGE compatible labeling of surface proteins on vital cells *in vitro* and *in vivo*, *Proteomics*, **6**, 579–585.

46. Heinemeyer J., Scheibe B., Schmitz U. K., Braun H. P. (2009) Blue native DIGE as a tool for comparative analyses of protein complexes. *J. Proteomic.*, **3**, 539–544.

47. Bisle B., Schmidt A., Scheibe B., Klein C., Tebbe A., Kellermann J., Siedler F., Pfeiffer F., Lottspeich F., Oesterhelt D. (2006) Quantitative profiling of the membrane proteome in a halophilicarchaeon. *Mol. Cell Proteomic.*, **5**, 1543–1558.

48. Hurd T. R., Prime T., Harbour M. E., Lilley K. S., Murphy M. P. (2007) Detection of reactive oxygen species-sensitive thiol proteins by redox difference gel electrophoresis. Implications for mitochondrial redox signaling. *Biol. Chem.*, **282**, 22040–22051.

Chapter 3

Mass Spectrometry Utilizing Isotope-Coded Affinity Tag Reagents

Naoyuki Yamada
Fundamental Technology Labs, Institute for Innovation, Ajinomoto Co., Inc., 1-1 Suzuki-cho, Kawasaki-ku, Kawasaki-shi 210-8681, Japan
naoyuki_yamada@ajinomoto.com

3.1 Introduction

Applications of mass spectrometry (MS) have expanded from the identification of proteins for proteome profiling to the quantitative comparison of proteome changes through various biological and physiological processes. Stable isotope labeling for protein quantitation is the most suitable method for use with MS. Various chemical derivatization methods have been developed for the incorporation of stable isotope tags onto peptides at their N- or C-terminals or on specific amino acids residues, such as cysteine, lysine, and tyrosine. One of these methods, the isotope-coded affinity tag (ICAT) technique, was developed to support comprehensive and quantitative protein expression profiling using reagents to chemically label specific cysteine residues with an affinity tag [1].

Quantitative Proteome Analysis: Methods and Applications
Edited by Kazuhiro Imai and Sam F. Y. Li

ISBN 978-981-4316-51-4 (Hardcover), 978-981-4316-98-9 (eBook)
www.panstanford.com

This approach can be used to quantify relative protein expression levels based on the relative intensity of the heavy and light forms of ICAT-labeled cysteines containing tryptic peptides as measured by tandem MS. Cysteine-specific modification and affinity purification achieves a reduction in proteome complexity. The ICAT technique has been widely applied to quantitative proteome analysis of a variety of sample types, such as cultured cells, tissues, and body fluids. Subsequently, a second generation of ICAT reagent was developed that contains an acid-cleavable moiety for removing the affinity tag. There are numerous articles and reviews [2–8] on the methodology and applications of ICAT technology, summarized here.

3.2 ICAT Reagents

ICAT reagents were developed for quantitation of protein expression levels through isotope labeling, and simultaneously achieve a reduction in proteome sample complexity. Two types of ICAT reagents have been developed, described below.

3.2.1 Original ICAT Reagent (First-Generation)

The original ICAT (first-generation) method was developed by Aebersold *et al.* [1] and consists of a thiolate reactive group for selective labeling of cysteinyl residues of proteins, an isotopic linker that incorporated a stable isotope (hydrogen or deuterium), and a biotin tag moiety for selective isolation of ICAT-labeled peptides (Fig. 3.1a). There are two forms of isotopic linkers, containing either eight ^{1}H atoms (light form, no deuterium) or eight ^{2}H (d8) atoms (heavy form), with an 8 Da mass difference between them.

The ICAT approach involves labeling two protein mixtures with the isotopically light and heavy ICAT reagents for comparison of their relative quantities (Fig. 3.2). The ICAT reagents are covalently bonded to the cysteine residues of the proteins. Proteins prepared from a control sample are treated with the light ICAT reagent, while proteins from the test sample are treated with the heavy reagent. After labeling, the two reacted protein mixtures are combined, digested with a protease such as trypsin, and the peptides are fractionated by cation exchange chromatography to remove several excess reagents such as residual tris(2-carboxyethyl)phosphine hydrochloride (TCEP), a reducing reagent. The cysteine-containing

ICAT-labeled peptides are separated from the other peptides using avidin affinity chromatography. In this step, only the cysteine-containing peptides are selected, effectively reducing the complexity of the proteome. The purified ICAT-labeled peptides are analyzed by high performance liquid chromatography–tandem MS (LC–MS/MS). The eluted peptides are ionized by electrospray ionization (ESI) and specific ions are selected for collision-induced dissociation (CID) in a data-dependent scan. The ICAT-labeled peptides are identified using the CID spectra and a sequence database search using a search engine such as Mascot (Matrix Science, Boston, MA) or SEQUEST (Thermo Scientific, San Jose, CA). Each pair of light and heavy ICAT-labeled peptides has an 8 Da mass difference in their mass measurements. The identified peptides are quantified by integrating the areas under the peaks of the extracted ion chromatograms for each isotope-coded peptide pair (light and heavy). The relative quantification is determined by the ratio of the peak areas of the peptide signal intensities for the test sample vs. the control sample.

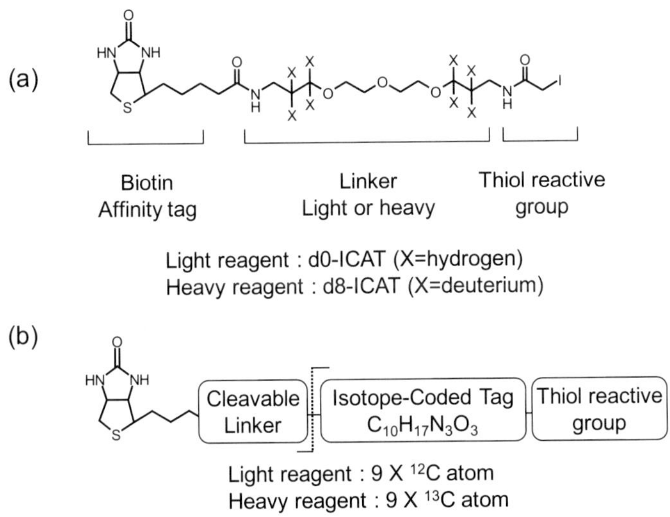

Figure 3.1 Structures of the (a) original and (b) cleavable isotope-coded affinity tag (ICAT) reagents. A biotin tag is used for affinity isolation of ICAT-labeled peptides. The linker for the heavy reagent incorporates a stable isotope; the original ICAT reagent uses eight deuterium atoms to replace eight ^{1}H atoms, and the cleavable ICAT reagent uses nine ^{13}C atoms to replace nine ^{12}C atoms. A thiol reactive group covalently and specifically binds to the cysteine residues of the proteins.

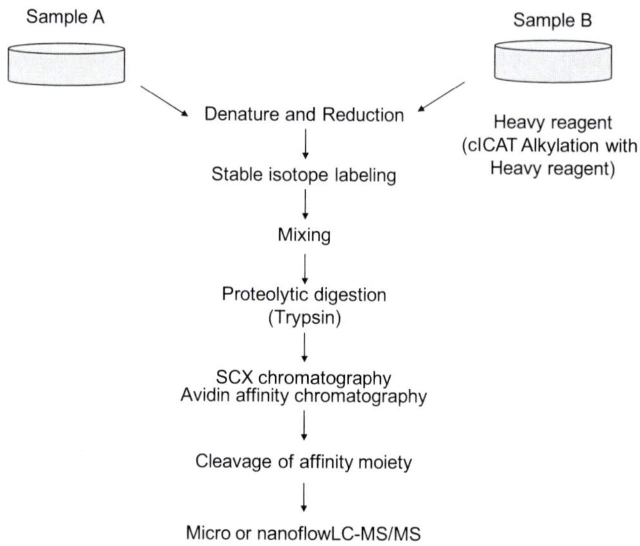

Figure 3.2 Workflow for quantitative analysis of proteome proteins using ICAT reagents.

The original ICAT approach has several limitations: (1) Incorporation of deuterium affects the physicochemical properties, in that the d8-ICAT-labeled peptides tend to elute before the d0-ICAT-labeled peptides in reverse-phase chromatography, which makes it difficult to quantitatively compare a single moment in time; (2) the biotin affinity tag is bulky and causes tag-related fragmentation of ICAT-labeled peptides in CID; (3) the rather hydrophobic biotin tag causes the peptides to elute within a fairly small time window; (4) potential isobaric ambiguity exists between the two ICAT-labeled peptides and the methionine oxidative peptides, due to the 8 Da mass difference between the heavy and light tag.

3.2.2 Cleavable ICAT (Second-Generation)

The second-generation ICAT (cleavable ICAT) approach incorporates an acid cleavable bond, which allows the removal of the large affinity tag moiety prior to MS measurements, and substitutes ^{13}C for ^{12}C in the heavy isotope-labeled reagent rather than deuterium for hydrogen (Fig. 3.1b). The cleavable ICAT reagent contains nine ^{13}C atoms in the heavy reagent instead of the eight deuterium atoms of the original ICAT reagent [9]. The polyethylene glycol linker of the original ICAT reagent was replaced with an acid-cleavable linker. An acid-cleavable isotope-coded affinity tag reagent (ICAT) with an undisclosed structure is available from the manufacturer (AB SCIEX, Foster City, CA).

The biotin tag of the cleavable ICAT reagent is removed by acid cleavage after avidin affinity chromatography. Following the avidin elution step, the cleavable ICAT-labeled peptides are evaporated and reconstituted in concentrated trifluoroacetic acid (TFA) to cleave the biotin moiety of the affinity tag from the peptides. This cleavage reduces the overall mass of the tag on the peptides and improves the peptide fragmentation efficiency of CID for sequence identification. After the cleavage of the avidin tag, the TFA is removed by evaporation. The purified cleaved ICAT-labeled peptides are analyzed by LC–MS/MS in the same manner as the original ICAT method for identification and relative quantification of the labeled peptides.

Cleavable ICAT uses ^{13}C atoms instead of deuterium to prepare the heavy labeled reagent, and for the cleavable ICAT reagent, the mass difference between the light and heavy forms is 9 Da. One of the limitations of the first-generation ICAT reagents was that the d0- and d8-labeled peptides had slightly different elution profiles in reverse-phase high-performance liquid chromatography (HPLC) separation due to the isotope effect. The cleavable ICAT reagent removes this limitation by incorporation of ^{13}C atoms instead of deuterium in the heavy reagent [10].

3.3 Methods

It is recommended that the manufacturer's instructions, cautionary statement, and data sheet be reviewed before using the ICAT reagent. The detailed cleavable-ICAT protocol and trouble-shooting methods have been reported by Shiio and Aebersold [8].

3.3.1 Denaturing and Reducing the Proteins

The protein concentration of the sample is determined using a Bradford assay prior to ICAT labeling. Eighty microliters of the denaturing buffer is added to test tubes containing 100 μg of the control sample and test sample. The protein samples are reduced by adding 2 μL of 50 mM TCEP (Sigma-Aldrich, St. Louis, MO) to both the control and test sample tubes. After vortexing to mix and spinning, the samples are boiled for 10 min.

3.3.2 Labeling with Cleavable ICAT Reagents

Twenty microliters of acetonitrile is added to vials containing Cleavable ICAT Light and Heavy Reagents (AB SCIEX, Foster City, CA). The reduced protein samples are transferred into the vials containing the cleavable ICAT light and heavy reagents and are then mixed and spun. The samples are incubated for 2 h at 30°C in the dark.

3.3.3 Separation of Labeled Proteins by SDS-PAGE and In-Gel Digestion by Trypsin

As an optional step to reduce the protein complexity, the ICAT-labeled proteins can be separated by sodium dodecyl sulfate polyacrylamide gel electrophoresis (SDS-PAGE). The labeled samples are combined into one tube and evaporated using a SpeedVac centrifugal vacuum concentrator (Thermo Scientific, San Jose, CA). The samples are then dissolved in 100 μL SDS-PAGE buffer and separated. The gel is stained with Coomassie blue and cut horizontally into several bands, and each band is subjected to gel trypsin digestion. Trypsin is added such that the ratio of the enzyme to the protein sample is 1:50 (wt/wt), and the samples are incubated at 37°C for 12–16 h.

3.3.4 Cation–Exchange Chromatography of the Cleavable ICAT–Labeled Peptides

Strong cation exchange (SCX) chromatography is used to remove excess ICAT reagents. One hundred micrograms of the peptide mixture is diluted by adding 2 mL Cation Exchange Buffer (5 mM KH_2PO_4, 25% acetonitrile, pH 2.5–3.0; AB SCIEX) and loaded onto an SCX cartridge equilibrated with 2 mL of the cation exchange buffer. The SCX cartridge is washed with the cation exchange buffer and loaded to elute the cleavable ICAT–labeled peptide using Cation Exchange Buffer-Elute (5 mM KH_2PO_4, 25% acetonitrile, 350 mM KCl, pH 2.5–3.0; AB SCIEX). After the peptide elution, the SCX cartridge is washed by injecting 1 mL of Cation Exchange Buffer-Clean (AB SCIEX).

To further reduce the protein complexity, the cleavable ICAT-labeled peptides can be fractionated using a high-resolution cation exchange column such as a polysulfoethyl A column (PolyLC, Inc., Columbia, MD) under gradient elution conditions.

3.3.5 Avidin-Affinity Chromatography and Cleavage of Biotin Moiety

The SCX-eluted fractions are neutralized to pH 8 using 100 mM NH_4HCO_3, pH 9.5 or according to the manufacturer's instructions as necessary to bring the pH up to 8.0. Each neutralized peptide fraction is separately purified using the avidin bead supplied with the cleavable ICAT reagent kit. The SCX-eluted fractions are then loaded onto the avidin cartridge. To remove the non-labeled peptides and nonspecific peptides, the avidin cartridge is washed and the cleavable ICAT-labeled peptides are eluted with Affinity Buffer–Elute (0.4% TFA in 30% acetonitrile; AB SCIEX).

Each avidin chromatography-eluted fraction is evaporated to dryness in a centrifugal vacuum concentrator. The biotin moiety is then cleaved under acidic conditions by adding 95 μL of cleaving reagent A and cleaving reagent B (AB SCIEX) and incubating for 2 h at 37°C. The cleaved sample is evaporated to dryness in a centrifugal vacuum concentrator for 30–60 min, and the biotin-cleaved peptides are reconstituted in 10–50 μL 0.1% TFA in 10% acetonitrile or 0.2% acetic acid–1% acetonitrile for MS.

3.3.6 Mass Spectrometry

MS data are collected using two different platforms: ESI MS with on-line nano- or microflow HPLC analysis and matrix-assisted laser desorption ionization (MALDI) MS with off-line separation using nano- or microflow HPLC fractionation.

The peptide samples are subjected to LC–MS/MS analysis using an ion trap, quadrupole time-of-flight (Q-TOF), TOF/TOF, or orbitrap type instrument equipped with a nanoelectrospray source, operating in positive ion mode. Typical HPLC conditions are a linear gradient using an octadecyl silane (ODS) reverse-phase column running a mobile phase consisting of 0.1% formic acid and acetonitrile. Generally, MS and CID data are collected continuously using a data-dependent scanning method. Information-dependent acquisition is used to select the most abundant precursor ions excluding singly charged ions for CID and the dynamic exclusion method is used to prevent the selection of the same precursor ions.

3.3.7 Sequence Database Search for Identification and Protein Quantitation

After analysis by nano- or microflow LC–MS/MS, the tandem mass spectra data are searched against a protein sequence database using search software such as SEQUEST, Mascot, or Pro ICAT (AB SCIEX). Quantitation is based on the relative intensities of the extracted ion chromatograms for precursor ions. Pro ICAT software incorporates the LC–MS Reconstruct quantitation algorithm to conduct relative quantitation of proteins. Mascot Distiller software, including the Search Toolbox and Quantitation Toolbox, can conduct peak picking, sequence searches, and relative quantitation.

References

1. Gygi, S.P. Rist, B. Gerber, S.A. Turecek, F. Gelb, M.H., and Aebersold, R. (1999) Quantitative analysis of complex protein mixtures using isotope-coded affinity tags. *Nat. Biotechnol.*, **17**, pp. 994–999.
2. Smolka, M.B. Zhou, H. Purkayastha, S., and Aebersold, R. (2001) Optimization of the isotope-coded affinity tag-labeling procedure for quantitative proteome analysis. *Anal. Biochem.*, **297**, pp. 25–31.
3. Gygi, S.P. Rist, B. Griffin, T.J. Eng, J., and Aebersold, R. (2002) Proteome analysis of low-abundance proteins using multidimensional chromatography and isotope-coded affinity tags. *J. Proteome. Res.*, **1**, pp. 47–54.
4. Goshe, M.B., and Smith, R.D. (2003) Stable isotope-coded proteomic mass spectrometry. *Curr. Opin. Biotechnol.*, **14**, pp. 101–109.
5. Hansen, K.C. Schmitt-Ulms, G. Chalkley, R.J. Hirsch, J. Baldwin, M.A., and Burlingame, A.L. (2003) Mass spectrometric analysis of protein mixtures at low levels using cleavable ^{13}C-isotope-coded affinity tag and multidimensional chromatography. *Mol. Cell. Proteomics.*, **2**, pp. 299–314.
6. Parker, K.C. Patterson, D. Williamson, B. Marchese, J. Graber, A. He, F. Jacobson, A. Juhasz, P., and Martin, S. (2004) Depth of proteome issues: a yeast isotope-coded affinity tag reagent study. *Mol. Cell. Proteomics.*, **3**, pp. 625–659.

7. Yi, E.C. Li, X.J. Cooke, K. Lee, H. Raught, B. Page, A. Aneliunas, V. Hieter, P. Goodlett, D.R., and Aebersold, R. (2005) Increased quantitative proteome coverage with $^{13}C/^{12}C$-based, acid-cleavable isotope-coded affinity tag reagent and modified data acquisition scheme. *Proteomics,* **5**, pp. 380–387.

8. Shiio, Y., and Aebersold, R. (2006) Quantitative proteome analysis using isotope-coded affinity tags and mass spectrometry. *Nat. Protoc.,* **1**, pp. 139–145.

9. Li, J. Steen, H., and Gygi, S.P. (2003) Protein profiling with cleavable isotope-coded affinity tag (cICAT) reagents: the yeast salinity stress response. *Mol. Cell Proteomics.,* **2**, pp. 1198–1204.

10. Zhang, R., and Regnier, F.E. (2002) Minimizing resolution of isotopically coded peptides in comparative proteomics. *J. Proteome. Res.,* **1**, pp. 139–147.

Chapter 4

Proteomic Analyses of Post-Translational Modifications

Wei-Chi Ku and Yasushi Ishihama
Graduate School of Pharmaceutical Sciences, Kyoto University, Kyoto 606-8501, Japan
yishiham@pharm.kyoto-u.ac.jp

Post-translational modifications (PTMs) regulate the activity and macromolecular interactions of many proteins. To date over 300 PTMs have been described, and many of them play essential roles in the pathogenesis of diseases. With recent advances in biological mass spectrometry (MS), comprehensive and accurate measurements of PTMs at the proteome scale has become possible, allowing a better understanding of cellular processes and disease mechanisms. Here we review recent proteomics methods for detection of some of the biologically important PTMs, including acetylation, glycosylation, *S*-nitrosylation, phosphorylation, and ubiquitylation.

4.1 Introduction

Post-translational modification refers to the chemical modification of specific amino acid residue(s) of a protein after its translation. PTMs

Quantitative Proteome Analysis: Methods and Applications
Edited by Kazuhiro Imai and Sam F. Y. Li

ISBN 978-981-4316-51-4 (Hardcover), 978-981-4316-98-9 (eBook)
www.panstanford.com

affect the protein's properties, such as charge and conformation, resulting in changes of activity, binding affinity to other proteins, localization, and turnover. For example, signal transduction cascades usually operate through a series of protein phosphorylation/dephosphorylation events [8]. Recent evidence has shown that PTMs not only affect protein activity individually, but also act in combination to modulate molecular interactions via multi-site modification, e.g., the "histone code" [10]. Therefore, systematic investigation of PTMs on a large scale is important to understand their role in cellular processes.

Traditionally, each PTM of interest has been studied at a specific amino acid residue in an individual protein by means of laborious biochemical approaches [93]. However, with the public release of complete human genome/proteome sequences and recent advances in mass spectrometry (MS)-based proteomics, it has become possible to identify PTM substrates, to map the PTM sites, and to quantitatively measure the dynamics of PTMs [3, 24, 30–31, 48, 93]. To date, more than 79, 171 experimentally proven or potential PTM sites in the human proteome have been recorded in the UniProt (http://www.uniprot.org) database (based on the UniProtKB/Swiss-Prot Release 2010_09). Table 4.1 summarizes some of the PTMs studied by means of MS-based proteomic approaches. A comprehensive list of post-translational modifications can be found in the UNIMOD database (http://www.unimod.org) and the Association of Biomolecular Resources Facilities (http://www.abrf.org).

4.2 Proteomic Approaches for PTM Studies

Although modern MS is a powerful and sensitive tool for proteomics, PTM studies still face many challenges [24, 30–31, 47, 93]. The relative low abundance of PTM proteins compared with unmodified proteins, the temporal dynamics of PTMs, and the spatial distributions of PTM proteins in subcellular organelles make direct measurement of PTMs with MS impossible. Therefore, enrichment or purification of PTM proteins of interest is important. Immunoaffinity purification using specific antibodies is the most common approach to enrich PTM proteins/peptides. The advantage of immunoaffinity purification is that sample complexity is greatly reduced for downstream LC-MS/MS analyses. However, high-quality

Table 4.1 Selected post-translational modifications commonly studied by means of proteomics approaches [24, 30–31, 47, 93]

Post-translational modification	Site of modification*	Change in mass (Da)	Biological functions
Acetylation	K (N-terminal, S, T)	42	Protein stability, protein–DNA interaction, regulation of enzyme activity.
Acylation Palmitoyl Farnesylation Myristoyl	 C (K, S, T) C K, R, N-terminal, G	 238 204 210	Cellular localization, protein–protein interaction, membrane tethering
Glycosylation N-linked O-linked	 N S, T	 >800 203, >800	Protein stability, protein solubility, secretion, cell-cell recognition
Methylation	K, R (D, E, H, P, C-terminal)	14, 28, 42	Regulation of gene expression, protein activity, protein–protein interaction, chromatin dynamics, protein–DNA interaction
S-Nitrosylation	C	29	Protein activity, inflammation response, oxidative protection, protein–protein interaction, transcription regulation
Phosphorylation	S, T, Y (H, R)	80	Protein activity, signaling, protein–protein interaction
Ubiquitylation	K (N-terminal)	>1000	Protein stability, signaling, transcription regulation

*Amino acid residues within brackets denote less common sites of modification.

antibodies are not always available for PTMs of interest. Based on the chemical properties of individual PTMs, alternative enrichment strategies may be considered (reviewed in Ref. [93]).

In addition to the enrichment of PTM proteins/peptides, further separation methods are usually applied, e.g., one- or two-dimensional gel electrophoresis for proteins, and ion exchange chromatography and isoelectric focusing for peptides. Alternatively, subcellular fractionation of biological samples also helps to reduce sample complexity and provides additional localization information.

Here, we review some commonly studied PTMs and MS-based proteomics approaches for these PTM studies.

4.2.1 Acetylation

Protein lysine acetylation, which is catalyzed by acetyltransferases using acetyl-CoA as the co-factor, was historically recognized to regulate gene transcription mainly by acting on histone proteins in the nucleus. The recent discovery of lysine acetylation of non-histone proteins, however, suggests a role of lysine acetylation in regulating cellular metabolism outside the nucleus [7, 92]. The reversible lysine acetylation of histones and nonhistone proteins has been shown to regulate a variety of physiological/pathological processes, such as aging, cancer, metabolism, inflammation, and cardiovascular protection [7]. Lysine acetylation, furthermore, shows cross talk with other PTMs such as protein phosphorylation, as exemplified by the DNA-damage-induced p53 activation signaling cascade [86].

Although recent advances in MS-based proteomics allow PTMs to be studied at the whole-proteome scale, e.g., phosphorylation [59], the small difference between two lysine PTMs, i.e., acetylation (42.011 Da increment) and tri-methylation (42.047 Da increment), impedes unambiguous site determination of lysine acetylation. With the availability of high-resolution mass spectrometers, such as the LTQ-Orbitrap, and specific antibodies to acetyl-lysine, effective purification and accurate measurement of lysine acetylation have become possible [34]. In a recent comprehensive proteomic survey on protein lysine acetylation, ~3500 acetylation sites in ~1700 acetylated proteins were identified in human cells, suggesting that lysine acetylation is a ubiquitous PTM, similar to phosphorylation [6]. The identification of protein acetylation in central metabolic enzymes in bacteria [80, 88, 91] and mammalian cells [6, 34, 67,

92] also demonstrates that protein acetylation is an evolutionarily conserved PTM to regulate cellular metabolism.

Alternatively, protein acetylation can also be studied by metabolic labeling with acetyl-CoA analogs [87] or affinity probes [27]. However, there is potential for off-target labeling with these chemical analogs or probes. This is because acetyl-CoA is required for many enzymes in addition to acetyltransferases. Furthermore, differential selectivity of chemical analogs for different acetyltransferases may cause biased labeling of a subset of protein substrates. Future development of "universal" chemical probes or other chemical derivations may provide tools that would be complementary to antibody-affinity purification for unbiased and comprehensive identification of acetylated proteins.

4.2.2 Glycosylation

Protein glycosylation plays key functions in many biological processes, including cell-to-cell recognition, membrane fusion, enzymatic activities, coordination of immune functions, and protein–protein interactions [13, 17]. Alteration of glycan structures on the glycoprotein or aberrant glycoprotein expression also correlates to the pathogenesis of many diseases, e.g., cancer and inflammation [12]. The two most common protein glycosylations are N-linked glycans, which attach to Asn residues in the motif Asn-X-(Ser/Thr) via *N*-acetylglucosamine, and O-linked glycans attached to Ser or Thr residues [54]. Other minor protein glycosylations on Cys, Lys, and Trp are also found in cells [25, 64].

Although carbohydrates on glycoproteins have been difficult to analyze in the past due to their structural complexity, recent development of MS-based proteomic approaches has allowed the identification of glycosylation sites and provided structural information for glycans [21]. For example, the glycoproteins or trypsin-digested glycopeptides are first enriched by multi-lectin affinity chromatography, and glycans are enzymatically or chemically released from the protein/peptide counterparts [21]. The glycoproteins can be chemically or enzymatically labeled with isotopic coded affinity tags (ICAT) [90] or ^{18}O [32] to perform quantitative proteomics studies. The released glycans can be separately analyzed to obtain structural information [52, 89]. At present, enormous efforts are being made to develop more sensitive MS methods for large-scale glycoproteomics studies.

4.2.3 *S*-Nitrosylation

Protein *S*-nitrosylation refers to the reversible covalent modification of the thiol group of cysteine residues by nitric oxide (NO). Protein *S*-nitrosylation plays an important role in NO-related and redox signaling pathways in many systems, from plants to mammalians [23, 43].

Due to the labile nature of the S–NO bond and the low abundance of endogenously *S*-nitrosylated proteins *in vivo*, however, it is challenging to identify *S*-nitrosylated proteins and *S*-nitrosylation sites, especially by MS-based proteomics approaches [33, 81]. In contrast to measuring the total amount of *S*-nitrosothiols by colorimetric or chemiluminescence detection (reviewed in Ref. [50]), the biotin switch method, which was first introduced by Jaffrey *et al.*, is the only viable MS-based method for the identification of protein *S*-nitrosylation at the proteomic level [29]. To date, over 300 proteins have been reported to be *S*-nitrosylated or regulated by *S*-nitrosylation, using the biotin switch method coupled with MS detection [16, 42, 68].

In the classic biotin-switch method as originally developed by Jaffrey *et al.* [29] (Fig. 4.1), free cysteines are first blocked by a thiol-reactive reagent, monomethyl thiosulphonate (MMTS), via disulfide bond formation. S–NO bonds of *S*-nitrosylated cysteine are then reduced with ascorbate. It should be noted that the reducing power of ascorbate is not strong enough to break disulfide linkages; therefore, specific reduction of the *S*-nitrosylated cysteine can be achieved. After chemical substitution with a biotin-containing affinity molecule, *N*-[6-(biotinamido)hexyl]-3′-(2′-pyridyldithio) propionamide (biotin-HPDP), followed by further biotin affinity purification, the biotinylated proteins are identified by LC-MS/MS.

Although the classic biotin-switch method has been widely applied for *S*-nitrosylation studies, some technical issues were raised [45]. For example, the disulfide bonds in proteins may hamper efficient trypsin digestion and downstream peptide identification. In addition, the decomposition of biotin-HPDP may produce a side reaction with free thiols, thereby introducing false-positive signals via disulfide interchange [15]. To overcome these problems, many technological improvements have been reported. By using an *S*-alkylating strategy to block free cysteines and to label *S*-nitrosylated cysteine, for example, disulfide interchange can be minimized and the endogenous disulfide bonds can be further reduced, allowing better digestion efficiency and identification confidence [2, 4]. In a recent study, Forrester *et al.* used a combination of a synthetic thiol-reactive resin and the iTRAQ labeling method to conduct a quantitative proteomics examination of *S*-nitrosylation [16]. For more comprehensive discussion of *S*-nitrosylation studies employing the biotin switch method, readers are referred to several recent review articles [15, 44, 49, 78].

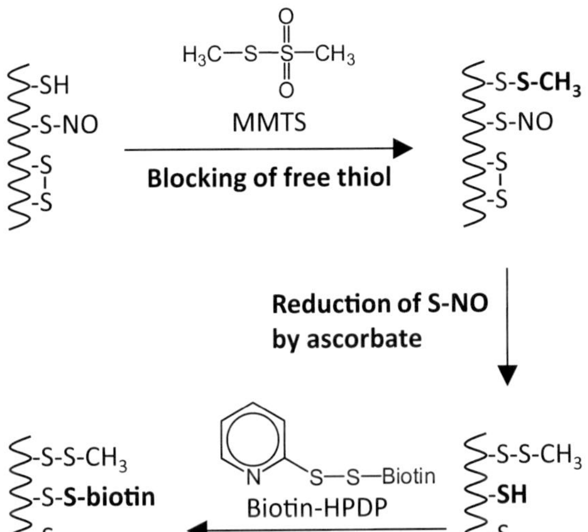

Figure 4.1 Chemical derivation of *S*-nitrosylated cysteine by the biotin switch method. Adapted from Ref. [29].

4.2.4 Phosphorylation

Protein phosphorylation, which occurs on Ser, Thr, or Tyr, is also a reversible PTM catalyzed by kinases and phosphatases. Phosphorylation is perhaps one of the most important PTMs in cells. One-third of all proteins in a eukaryotic cell can undergo phosphorylation modifications [48]. Protein phosphorylation participates in the regulation of protein activities, signaling transduction, protein–protein interaction, and control of transcription [26].

MS-based approaches have been extensively applied to the study of protein phosphorylation at the proteome scale, i.e., the phosphoproteome [11, 46, 48, 76]. For example, Rush *et al.* used antibodies specifically targeting phospho-tyrosine to perform immunoaffinity profiling of tyrosine phosphorylation in cancer cells [65]. Due to the unavailability of phospho-serine and phospho-threonine specific antibodies, alternatively, several chemical methods for phosphoproteins enrichment have been reported, such as immobilized metal ion affinity chromatography (IMAC) [70], metal oxide chromatography [28, 38], and β-elimination followed by Michael addition [18, 58, 77]. Modifications based on the above-mentioned strategies have been successfully applied for comprehensive phosphoproteomic studies [19, 37, 41, 63, 72, 79]. Recent developments and applications are discussed in the later part of this chapter.

4.2.5 Ubiquitylation

Protein ubiquitylation is the addition of ubiquitin, which is a small protein consisting of 76 amino acid residues, via its C-terminal residue (Gly76) to a substrate lysine residue by ubiquitin-ligase enzymes [22]. Protein ubiquitylation was first found to target proteins for proteasomal degradation, thereby playing an important role in the regulation of protein turnover in cells [22]. The later discovery of many ubiquitin-like modifiers, such as NEDD8 or SUMO, further extended the non-degradative functions of protein ubiquitylation to a variety of biological processes, such as DNA repair, transcription, endocytosis, sorting, and signal transduction (reviewed in Ref. [83]).

Although it is inherently difficult to study protein ubiquitylation due to the rapid turnover of ubiquitylated proteins, several proteomics approaches have been developed (reviewed in Ref. [35]). For example, Peng *et al.* introduced a recombinant 6xHis ubiquitin gene into yeast cells, and proteins labeled with 6xHis-ubiquitin were purified by His affinity chromatography. After trypsin digestion followed by strong cation exchange chromatography and reversed-phase LC-MS/MS, 1075 proteins were identified from yeasts [60]. Furthermore, 110 ubiquitylation sites present in 72 substrate proteins were also preciously determined by taking the advantage of a mass shift at lysine residue (+114.1 Da), which derived from di-glycine residues (Gly-Gly) of ubiquitin after trypsin digestion [60].

Matsumoto *et al.* performed immunoaffinity chromatography using a monoclonal antibody that specifically recognized ubiquitin-protein conjugates, but not free ubiquitin, to purify ubiquitylated proteins from human HEK 293T cells [51]. By varying the purification environment (native or denaturing conditions), target proteins in substantially different functional categories were identified. In total, 670 distinct proteins were identified; 345 proteins and 325 proteins were identified after denaturing and native purification, respectively [51]. The different subsets of ubiquitylated proteins identified under the two conditions may reflect potential regulation and quality control of different proteins, e.g., ribosomal proteins [51].

Although the signature mass shift (+114.1 Da) at di-glycine-containing lysine peptides from ubiquitylated protein has been widely applied to pin-point ubiquitylation sites, Nielsen *et al.* reported that an iodoacetamide-induced artifact, which mimics ubiquitylation signatures, may arise during sample preparation [57]. Therefore, because of the potential for false-positive identification, bona fide ubiquitylation sites should be verified by other biochemical approaches.

4.3 MS-Based Analyses of Protein Phosphorylation

4.3.1 Enrichment of Phosphorylated Proteins or Peptides

Phosphorylated proteins are usually characterized by MS after proteolytic digestion. Due to the lower ionization efficiency of phosphopeptides, efficient and specific enrichment of the phosphopeptides prior to MS analyses is important. As mentioned in Section 4.2.4, various enrichment strategies are currently available. Figure 4.2 gives an overview of these strategies. Here we focus on some commonly used phosphopeptide enrichment approaches. Readers who are interested in quantitative phosphoproteomics should refer to several recent reviews and studies [76, 82].

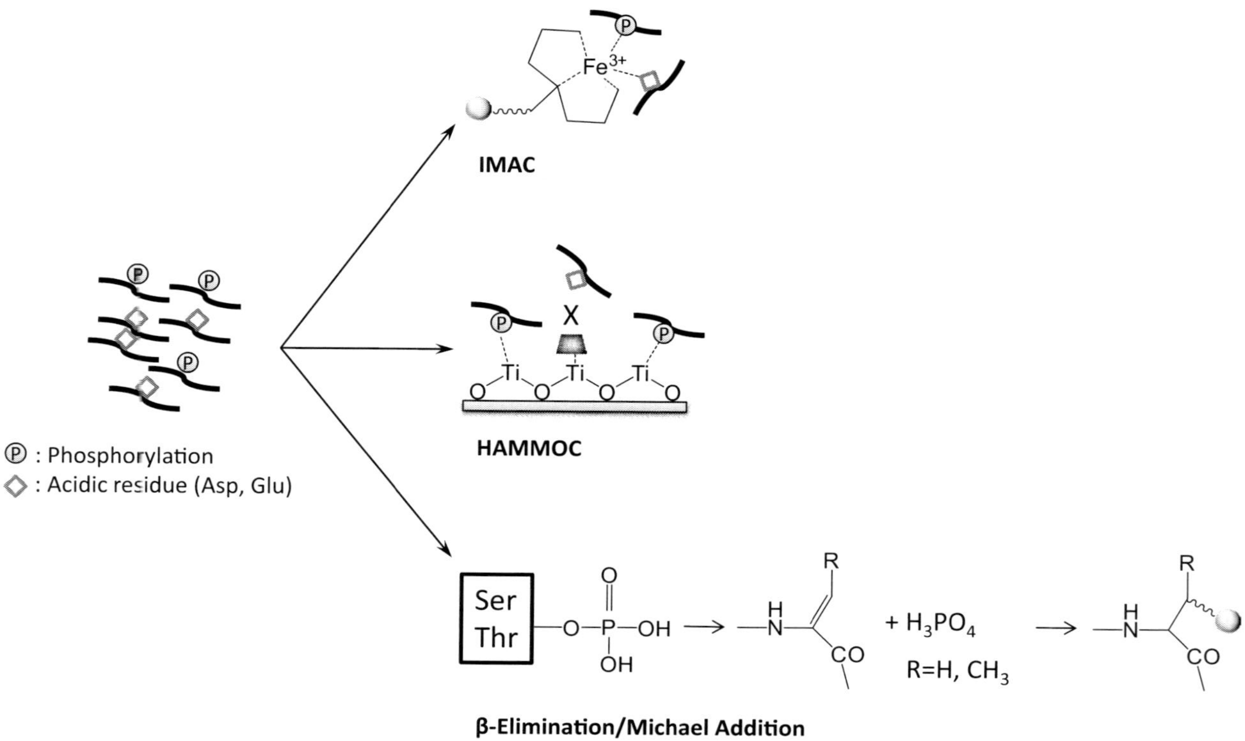

Figure 4.2 Representative strategies for phosphopeptide enrichment. Adapted from Ref. [76].

4.3.1.1 Immobilized metal ion affinity chromatography

Immobilized metal ion affinity chromatography is traditionally used for the affinity purification of His-tagged proteins by chelating metal ions (Fe^{3+}, Ga^{3+}, or Co^{2+}) to nitrilotriacetic acid (NTA) or iminodiacetic acid (IDA) coated beads. As first demonstrated by Neville *et al.* [56], phosphoproteins and phosphoamino acids also bind to IMAC resin [1], and so IMAC can be used for the enrichment of phosphopeptides.

Although IMAC has been extensively used for phosphopeptide enrichment, the high level of non-specific binding of nonphosphorylated peptides is a serious problem, especially in highly complex peptide samples. This non-specific binding is due to the interaction between IMAC resin and the carboxylic side chains of Asp and Glu in nonphosphorylated peptides. Several approaches have been developed to circumvent the problem. Ficarro *et al.* improved the enrichment specificity by using *O*-methyl esterification to block carboxyl groups of Asp and Glu [14]. Alternatively, the specificity of IMAC can also be increased by adjusting the pH to the range (2.7~3.0) in which carboxyl groups of Asp and Glu are protonated and lose their binding affinity to IMAC resins [62]. By carefully controlling the species and concentration of acids, specific and effective enrichment of phosphopeptides can be achieved [37, 79].

4.3.1.2 Hydroxy acid-modified metal oxide chromatography

TiO_2 beads were first used as an affinity agent for organophosphates, including phosphopeptides [28], and were subsequently used for phosphoproteomic studies [61]. In addition, ZrO_2 has been used for phosphopeptide enrichment, because of the high affinity of ZrO_2 for phosphates [38]. By including benzoic acid derivatives, e.g., 2,5-dihydroxybenzoic acid (DHB) and phthalic acid, in the sample loading buffer, Larsen *et al.* achieved effective removal of acidic non-phosphopeptides during phosphopeptide enrichment with TiO_2 and downstream MALDI-MS analyses. Although this protocol is proven effective, it is not applicable to LC-MS/MS analysis, because residual DHB interferes with peptide detection.

Our group recently developed a novel method for phosphopeptide enrichment using aliphatic hydroxy acid-modified metal oxide chromatography (HAMMOC) [72]. By the inclusion of aliphatic hydroxy acid, the HAMMOC system greatly reduces the non-specific binding of non-phosphopeptides [72]. Hydroxy acids can bind to metal oxide, e.g., TiO_2 and ZrO_2, by forming a cyclic chelate complex [75], whereas phosphate forms a bridging complex with two metal ions [9]. Therefore, there is a differential affinity for metal oxides in the order phosphopeptides > aliphatic hydroxy acid > nonphosphopeptides (strong to weak) [72]. By screening a series of hydroxy acids as sampling loading systems, we found that lactic acid most effectively suppressed the identification of nonphosphopeptides [72]. At the same time, the lactic acid can be easily removed by means of a C18 desalting procedure; therefore the HAMMOC system can be coupled with LC-MS/MS without difficulty. By using successive elution of phosphopeptides with secondary amines, such as piperidine and pyrrolidine, in addition to ammonium hydroxide, we were able to identify 3000~4000 phosphopeptides using HAMMOC from only 100 μg HeLa cytoplasmic extract as the starting material [39].

4.3.1.3 β-Elimination and Michael addition

Phosphopeptides can also be enriched by chemical derivatization. Phosphate groups on serine and threonine can be removed by alkaline treatment, i.e., β-elimination, to form dehydroalanine and dehydrobutyrine, respectively. When a sulfhydryl group containing a linker coupled with biotin or another affinity tag is added, a dithiol is formed in the peptide by Michael addition, allowing labeling and enrichment of phosphopeptides for LC-MS/MS analyses [18, 58]. However, there are some drawbacks to β-elimination/Michael addition strategies. A large amount of starting material is required due to the complicated chemical reaction steps, and, more importantly, β-elimination can occur on *O*-glycosylation, resulting in false identification of phosphopeptides.

Despite the drawbacks of phosphopeptide enrichment, β-elimination can be useful for the confirmation of phosphopeptides after purification by other strategies, as first demonstrated by Wolschin *et al.* [84]. We recently presented a novel strategy that consists of HAMMOC-based phosphopeptide enrichment followed by alkali-induced β-elimination, to improve the efficiency of phosphopeptide identification by MS, especially for multiply phosphorylated phosphopeptides [40]. By using this strategy, we have successfully identified 1649 unique phosphopeptides, including 325 multiply phosphorylated phosphopeptides, from 200 μg *Arabidopsis* cellular lysate [40].

4.3.2 Application of HAMMOC in Phosphoproteomic Studies

With the success of MS-based phosphoproteomics approaches, more than 20,000 and 8000 phosphorylation sites in the human and mouse proteome, respectively, have been documented in the UniProt-SwissProt database. In fact, as shown in Fig. 4.3, our group has found more than 111,000 phosphorylation sites in 14,627 human proteins by using the HAMMOC strategy (Ishihama *et al.*, unpublished data in April, 2013). When the UniProt entries are combined with our data, it is found that 72% of human proteins are subject to phosphorylation modification.

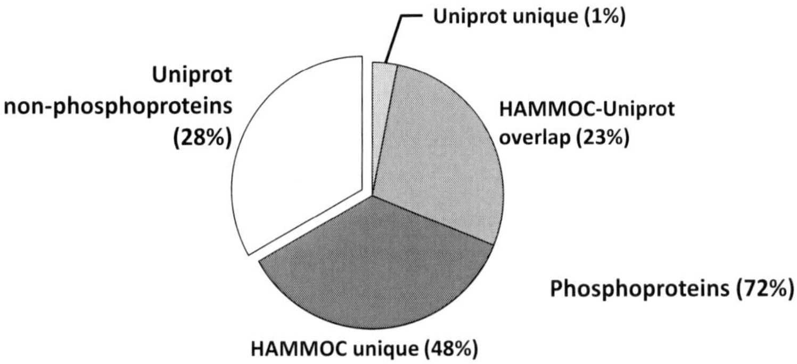

Figure 4.3 Nearly two-thirds of human proteins can be phosphorylated.

Besides mammalian cells, we also studied protein phosphorylation in other organisms using the HAMMOC strategy. For example, over 2000 and 6000 phosphorylation sites in *Arabidopsis* [40, 73] and rice [55], respectively, were identified by our group and colleagues. These data indicate that protein phosphorylation is an evolutionarily conserved PTM in plants and mammals.

4.3.3 Sequencing and Site-Determination of Phosphopeptides by MS

In modern proteomics studies, peptide sequencing with tandem MS is mostly performed by collision-induced dissociation (CID) [69]. However, sequencing of phosphopeptides by CID is challenging due to the loss of labile phosphate groups and the low ionization efficiency.

Several MS-based methods have been developed to improve the identification efficiency and site accuracy of phosphopeptides, such as neutral loss-triggered MS^3 [20], pseudo-MS^3 [66], electron capture dissociation (ECD) [36, 71], and electron transfer dissociation (ETD) [5, 53]. It should be noted that ECD and ETD are useful methods for the identification of multiply phosphorylated peptides, which are difficult to analyze by CID [74].

4.4 Concluding Remarks

PTM proteomics is an emerging field that has progressed rapidly in recent years. The continuous methodological developments in PTM detection, together with the development of increasingly sophisticated mass spectrometers, means that comprehensive mapping of individual PTMs is now feasible. The next important issue is to establish how PTMs crosstalk with each other to regulate cellular processes at the proteome scale, as exemplified by the multiplex PTM modification of histones, the so-called "histone code" [85]. We believe that comprehensive multiplex PTM studies will provide a better understanding of cellular processes and homeostasis in the future.

References

1. Andersson, L., and Porath, J. (1986). Isolation of phosphoproteins by immobilized metal (Fe^{3+}) affinity chromatography, *Anal. Biochem.*, **154**, pp. 250–254.

2. Camerini, S., Polci, M. L., Restuccia, U., Usuelli, V., Malgaroli, A., and Bachi, A. (2007). A novel approach to identify proteins modified by nitric oxide: the HIS-TAG switch method, *J. Proteome Res.*, **6**, pp. 3224–3231.

3. Cantin, G. T., and Yates, J. R., 3rd (2004). Strategies for shotgun identification of post-translational modifications by mass spectrometry, *J. Chromatogr. A*, **1053**, pp. 7–14.

4. Chen, Y. J., Ku, W. C., Lin, P. Y., Chou, H. C., Khoo, K. H., and Chen, Y. J. (2010). An *S*-alkylating labeling strategy for site-specific identification of the *S*-nitrosoproteome, *J. Proteome Res.*, **9**, pp. 6417–6439.

5. Chi, A., Huttenhower, C., Geer, L. Y., Coon, J. J., Syka, J. E., Bai, D. L., Shabanowitz, J., Burke, D. J., Troyanskaya, O. G., and Hunt, D. F. (2007).

Analysis of phosphorylation sites on proteins from Saccharomyces cerevisiae by electron transfer dissociation (ETD) mass spectrometry, *Proc. Natl. Acad. Sci. U. S. A.*, **104**, pp. 2193–2198.

6. Choudhary, C., Kumar, C., Gnad, F., Nielsen, M. L., Rehman, M., Walther, T. C., Olsen, J. V., and Mann, M. (2009). Lysine acetylation targets protein complexes and co-regulates major cellular functions, *Science*, **325**, pp. 834–840.
7. Close, P., Creppe, C., Gillard, M., Ladang, A., Chapelle, J. P., Nguyen, L., and Chariot, A. (2010). The emerging role of lysine acetylation of non-nuclear proteins, *Cell. Mol. Life Sci.*, **67**, pp. 1255–1264.
8. Cohen, P. (2000). The regulation of protein function by multisite phosphorylation—a 25 year update, *Trends Biochem. Sci.*, **25**, pp. 596-601.
9. Connor, P. A., and McQuillan, A. J. (1999). Phosphate adsorption onto TiO_2 from aqueous solutions: an *in situ* internal reflection infrared spectroscopic study, *Langmuir*, **15**, pp. 2916–2921.
10. Cosgrove, M. S., Boeke, J. D., and Wolberger, C. (2004). Regulated nucleosome mobility and the histone code, *Nat. Struct. Mol. Biol.*, **11**, pp. 1037–1043.
11. Dass, C. (2009). Recent developments in mass spectrometry analysis of phosphoproteomes, *Curr. Proteomics*, **6**, pp. 32–42.
12. Dube, D. H., and Bertozzi, C. R. (2005). Glycans in cancer and inflammation—potential for therapeutics and diagnostics, *Nat. Rev. Drug Discov.*, **4**, pp. 477–488.
13. Dwek, R. A. (1996). Glycobiology: toward understanding the function of sugars, *Chem. Rev.*, **96**, pp. 683–720.
14. Ficarro, S. B., McCleland, M. L., Stukenberg, P. T., Burke, D. J., Ross, M. M., Shabanowitz, J., Hunt, D. F., and White, F. M. (2002). Phosphoproteome analysis by mass spectrometry and its application to *Saccharomyces cerevisiae*, *Nat. Biotechnol.*, **20**, pp. 301–305.
15. Forrester, M. T., Foster, M. W., and Stamler, J. S. (2007). Assessment and application of the biotin switch technique for examining protein S-nitrosylation under conditions of pharmacologically induced oxidative stress, *J. Biol. Chem.*, **282**, pp. 13977–13983.
16. Forrester, M. T., Thompson, J. W., Foster, M. W., Nogueira, L., Moseley, M. A., and Stamler, J. S. (2009). Proteomic analysis of *S*-nitrosylation and denitrosylation by resin-assisted capture, *Nat. Biotech.*, **27**, pp. 557–559.

17. Freeze, H. H., and Aebi, M. (2005). Altered glycan structures: the molecular basis of congenital disorders of glycosylation, *Curr. Opin. Struct. Biol.*, **15**, pp. 490–498.

18. Goshe, M. B., Conrads, T. P., Panisko, E. A., Angell, N. H., Veenstra, T. D., and Smith, R. D. (2001). Phosphoprotein isotope-coded affinity tag approach for isolating and quantitating phosphopeptides in proteome-wide analyses, *Anal. Chem.*, **73**, pp. 2578–2586.

19. Goshe, M. B., Veenstra, T. D., Panisko, E. A., Conrads, T. P., Angell, N. H., and Smith, R. D. (2002). Phosphoprotein isotope-coded affinity tags: application to the enrichment and identification of low-abundance phosphoproteins, *Anal. Chem.*, **74**, pp. 607–616.

20. Gruhler, A., Olsen, J. V., Mohammed, S., Mortensen, P., Faergeman, N. J., Mann, M., and Jensen, O. N. (2005). Quantitative phosphoproteomics applied to the yeast pheromone signaling pathway, *Mol. Cell. Proteomics*, **4**, pp. 310–327.

21. Harvey, D. J. (2005). Proteomic analysis of glycosylation: structural determination of N- and O-linked glycans by mass spectrometry, *Expert Rev. Proteomics*, **2**, pp. 87–101.

22. Hershko, A., and Ciechanover, A. (1998). The ubiquitin system, *Annu. Rev. Biochem.*, **67**, pp. 425–479.

23. Hess, D. T., Matsumoto, A., Kim, S.-O., Marshall, H. E., and Stamler, J. S. (2005). Protein *S*-nitrosylation: purview and parameters, *Nat. Rev. Mol. Cell Biol.*, **6**, pp. 150–166.

24. Hoffman, M. D., Sniatynski, M. J., and Kast, J. (2008). Current approaches for global post-translational modification discovery and mass spectrometric analysis, *Anal. Chim. Acta*, **627**, pp. 50–61.

25. Hofsteenge, J., Muller, D. R., de Beer, T., Loffler, A., Richter, W. J., and Vliegenthart, J. F. (1994). New type of linkage between a carbohydrate and a protein: C-glycosylation of a specific tryptophan residue in human RNase Us, *Biochemistry*, **33**, pp. 13524–13530.

26. Hunter, T. (2000). Signaling—2000 and beyond, *Cell*, **100**, pp. 113–127.

27. Hwang, Y., Thompson, P. R., Wang, L., Jiang, L., Kelleher, N. L., and Cole, P. A. (2007). A selective chemical probe for coenzyme A-requiring enzymes, *Angew. Chem. Int. Ed. Engl.*, **46**, pp. 7621–7624.

28. Ikeguchi, Y., and Nakamura, H. (1997). Determination of organic phosphates by column-switching high performance anion-exchange chromatography using on-line preconcentration on titania, *Anal. Sci.*, **13**, pp. 479–483.

29. Jaffrey, S. R., Erdjument-Bromage, H., Ferris, C. D., Tempst, P., and Snyder, S. H. (2001). Protein *S*-nitrosylation: a physiological signal for neuronal nitric oxide, *Nat. Cell. Biol.*, **3**, pp. 193–197.

30. Jensen, O. N. (2004). Modification-specific proteomics: characterization of post-translational modifications by mass spectrometry, *Curr. Opin. Chem. Biol.*, **8**, pp. 33–41.

31. Jensen, O. N. (2006). Interpreting the protein language using proteomics, *Nat. Rev. Mol. Cell. Biol.*, **7**, pp. 391–403.

32. Kaji, H., Saito, H., Yamauchi, Y., Shinkawa, T., Taoka, M., Hirabayashi, J., Kasai, K., Takahashi, N., and Isobe, T. (2003). Lectin affinity capture, isotope-coded tagging and mass spectrometry to identify N-linked glycoproteins, *Nat. Biotechnol.*, **21**, pp. 667–672.

33. Kaneko, R., and Wada, Y. (2003). Decomposition of protein nitrosothiols in matrix-assisted laser desorption/ionization and electrospray ionization mass spectrometry, *J. Mass Spectrom.*, **38**, pp. 526–530.

34. Kim, S. C., Sprung, R., Chen, Y., Xu, Y., Ball, H., Pei, J., Cheng, T., Kho, Y., Xiao, H., Xiao, L., Grishin, N. V., White, M., Yang, X. J., and Zhao, Y. (2006). Substrate and functional diversity of lysine acetylation revealed by a proteomics survey, *Mol. Cell*, **23**, pp. 607–618.

35. Kirkpatrick, D. S., Denison, C., and Gygi, S. P. (2005). Weighing in on ubiquitin: the expanding role of mass-spectrometry-based proteomics, *Nat. Cell Biol.*, **7**, pp. 750–757.

36. Kleinnijenhuis, A. J., Kjeldsen, F., Kallipolitis, B., Haselmann, K. F., and Jensen, O. N. (2007). Analysis of histidine phosphorylation using tandem MS and ion-electron reactions, *Anal. Chem.*, **79**, pp. 7450–7456.

37. Kokubu, M., Ishihama, Y., Sato, T., Nagasu, T., and Oda, Y. (2005). Specificity of immobilized metal affinity-based IMAC/C18 tip enrichment of phosphopeptides for protein phosphorylation analysis, *Anal. Chem.*, **77**, pp. 5144–5154.

38. Kweon, H. K., and Hakansson, K. (2006). Selective zirconium dioxide-based enrichment of phosphorylated peptides for mass spectrometric analysis, *Anal. Chem.*, **78**, pp. 1743–1749.

39. Kyono, Y., Sugiyama, N., Imami, K., Tomita, M., and Ishihama, Y. (2008). Successive and selective release of phosphorylated peptides captured by hydroxy acid-modified metal oxide chromatography, *J. Proteome Res.*, **7**, pp. 4585–4593.

40. Kyono, Y., Sugiyama, N., Tomita, M., and Ishihama, Y. (2010). Chemical dephosphorylation for identification of multiply phosphorylated

peptides and phosphorylation site determination, *Rapid Commun. Mass Spectrom.*, **24**, pp. 2277–2282.

41. Larsen, M. R., Thingholm, T. E., Jensen, O. N., Roepstorff, P., and Jorgensen, T. J. (2005). Highly selective enrichment of phosphorylated peptides from peptide mixtures using titanium dioxide microcolumns, *Mol. Cell. Proteomics*, **4**, pp. 873–886.

42. Lefièvre, L., Chen, Y., Conner, S. J., Scott, J. L., Publicover, S. J., Ford, W. C. L., and Barratt, C. L. R. (2007). Human spermatozoa contain multiple targets for protein *S*-nitrosylation: an alternative mechanism of the modulation of sperm function by nitric oxide?, *Proteomics*, **7**, pp. 3066–3084.

43. Lindermayr, C., Saalbach, G., and Durner, J. (2005). Proteomic identification of *S*-nitrosylated proteins in *Arabidopsis*, *Plant Physiol.*, **137**, pp. 921–930.

44. Lindermayr, C., Sell, S., and Durner, J. (2009). Generation and detection of S-nitrosothiols, *Methods Mol. Biol.*, **476**, pp. 210–222.

45. López-Sánchez, L. M., Muntané, J., Mata, M. d. l., and Rodríguez-Ariza, A. (2009). Unraveling the *S*-nitrosoproteome: tools and strategies, *Proteomics*, **9**, pp. 808–818.

46. Macek, B., Mann, M., and Olsen, J. V. (2009). Global and site-specific quantitative phosphoproteomics: principles and applications, *Annu. Rev. Pharmacol. Toxicol.*, **49**, pp. 199–221.

47. Mann, M., and Jensen, O. N. (2003). Proteomic analysis of post-translational modifications, *Nat. Biotechnol.*, **21**, pp. 255–261.

48. Mann, M., Ong, S.-E., Grønborg, M., Steen, H., Jensen, O. N., and Pandey, A. (2002). Analysis of protein phosphorylation using mass spectrometry: deciphering the phosphoproteome, *Trends Biotechnol.*, **20**, pp. 261–268.

49. Mannick, J. B., and Schonhoff, C. M. (2008). Measurement of protein S-nitrosylation during cell signaling, *Methods Enzymol.*, **440**, pp. 231–242.

50. Martinez-Ruiz, A., and Lamas, S. (2004). *S*-nitrosylation: a potential new paradigm in signal transduction, *Cardiovasc. Res.*, **62**, pp. 43–52.

51. Matsumoto, M., Hatakeyama, S., Oyamada, K., Oda, Y., Nishimura, T., and Nakayama, K. I. (2005). Large-scale analysis of the human ubiquitin-related proteome, *Proteomics*, **5**, pp. 4145–4151.

52. Mechref, Y., and Novotny, M. V. (2002). Structural investigations of glycoconjugates at high sensitivity, *Chem. Rev.*, **102**, pp. 321–369.

53. Molina, H., Horn, D. M., Tang, N., Mathivanan, S., and Pandey, A. (2007). Global proteomic profiling of phosphopeptides using electron transfer dissociation tandem mass spectrometry, *Proc. Natl. Acad. Sci. U. S. A.*, **104**, pp. 2199–2204.

54. Morelle, W., Canis, K., Chirat, F., Faid, V., and Michalski, J. C. (2006). The use of mass spectrometry for the proteomic analysis of glycosylation, *Proteomics*, **6**, pp. 3993–4015.

55. Nakagami, H., Sugiyama, N., Mochida, K., Daudi, A., Yoshida, Y., Toyoda, T., Tomita, M., Ishihama, Y., and Shirasu, K. (2010). Large-scale comparative phosphoproteomics identifies conserved phosphorylation sites in plants, *Plant Physiol.*, **153**, pp. 1161–1174.

56. Neville, D. C., Rozanas, C. R., Price, E. M., Gruis, D. B., Verkman, A. S., and Townsend, R. R. (1997). Evidence for phosphorylation of serine 753 in CFTR using a novel metal-ion affinity resin and matrix-assisted laser desorption mass spectrometry, *Protein Sci.*, **6**, pp. 2436–2445.

57. Nielsen, M. L., Vermeulen, M., Bonaldi, T., Cox, J., Moroder, L., and Mann, M. (2008). Iodoacetamide-induced artifact mimics ubiquitination in mass spectrometry, *Nat. Methods*, **5**, pp. 459–460.

58. Oda, Y., Nagasu, T., and Chait, B. T. (2001). Enrichment analysis of phosphorylated proteins as a tool for probing the phosphoproteome, *Nat. Biotechnol.*, **19**, pp. 379–382.

59. Olsen, J. V., Blagoev, B., Gnad, F., Macek, B., Kumar, C., Mortensen, P., and Mann, M. (2006). Global, *in vivo*, and site-specific phosphorylation dynamics in signaling networks, *Cell*, **127**, pp. 635–648.

60. Peng, J., Schwartz, D., Elias, J. E., Thoreen, C. C., Cheng, D., Marsischky, G., Roelofs, J., Finley, D., and Gygi, S. P. (2003). A proteomics approach to understanding protein ubiquitination, *Nat. Biotechnol.*, **21**, pp. 921–926.

61. Pinkse, M. W., Uitto, P. M., Hilhorst, M. J., Ooms, B., and Heck, A. J. (2004). Selective isolation at the femtomole level of phosphopeptides from proteolytic digests using 2D-NanoLC-ESI-MS/MS and titanium oxide precolumns, *Anal. Chem.*, **76**, pp. 3935–3943.

62. Posewitz, M. C., and Tempst, P. (1999). Immobilized gallium(III) affinity chromatography of phosphopeptides, *Anal. Chem.*, **71**, pp. 2883–2892.

63. Qian, W. J., Goshe, M. B., Camp, D. G., 2nd, Yu, L. R., Tang, K., and Smith, R. D. (2003). Phosphoprotein isotope-coded solid-phase tag approach for enrichment and quantitative analysis of phosphopeptides from complex mixtures, *Anal. Chem.*, **75**, pp. 5441–5450.

64. Rajeswari, P., Natarajan, R., Nadler, J. L., Kumar, D., and Kalra, V. K. (1991). Glucose induces lipid peroxidation and inactivation of membrane-associated ion-transport enzymes in human erythrocytes *in vivo* and *in vitro*, *J. Cell Physiol.*, **149**, pp. 100–109.
65. Rush, J., Moritz, A., Lee, K. A., Guo, A., Goss, V. L., Spek, E. J., Zhang, H., Zha, X. M., Polakiewicz, R. D., and Comb, M. J. (2005). Immunoaffinity profiling of tyrosine phosphorylation in cancer cells, *Nat. Biotechnol.*, **23**, pp. 94–101.
66. Schroeder, M. J., Shabanowitz, J., Schwartz, J. C., Hunt, D. F., and Coon, J. J. (2004). A neutral loss activation method for improved phosphopeptide sequence analysis by quadrupole ion trap mass spectrometry, *Anal. Chem.*, **76**, pp. 3590–3598.
67. Schwer, B., Eckersdorff, M., Li, Y., Silva, J. C., Fermin, D., Kurtev, M. V., Giallourakis, C., Comb, M. J., Alt, F. W., and Lombard, D. B. (2009). Calorie restriction alters mitochondrial protein acetylation, *Aging Cell*, **8**, pp. 604–606.
68. Stamler, J. S., Lamas, S., and Fang, F. C. (2001). Nitrosylation: the prototypic redox-based signaling mechanism, *Cell*, **106**, pp. 675–683.
69. Steen, H., and Mann, M. (2004). The ABC's (and XYZ's) of peptide sequencing, *Nat. Rev. Mol. Cell Biol.*, **5**, pp. 699–711.
70. Stensballe, A., Andersen, S., and Jensen, O. N. (2001). Characterization of phosphoproteins from electrophoretic gels by nanoscale Fe(III) affinity chromatography with off-line mass spectrometry analysis, *Proteomics*, **1**, pp. 207–222.
71. Stensballe, A., Jensen, O. N., Olsen, J. V., Haselmann, K. F., and Zubarev, R. A. (2000). Electron capture dissociation of singly and multiply phosphorylated peptides, *Rapid Commun. Mass Spectrom.*, **14**, pp. 1793–1800.
72. Sugiyama, N., Masuda, T., Shinoda, K., Nakamura, A., Tomita, M., and Ishihama, Y. (2007). Phosphopeptide enrichment by aliphatic hydroxy acid-modified metal oxide chromatography for nano-LC-MS/MS in proteomics applications, *Mol. Cell. Proteomics*, **6**, pp. 1103–1109.
73. Sugiyama, N., Nakagami, H., Mochida, K., Daudi, A., Tomita, M., Shirasu, K., and Ishihama, Y. (2008). Large-scale phosphorylation mapping reveals the extent of tyrosine phosphorylation in Arabidopsis, *Mol. Syst. Biol.*, **4**, p. 193.
74. Swaney, D. L., Wenger, C. D., Thomson, J. A., and Coon, J. J. (2009). Human embryonic stem cell phosphoproteome revealed by electron transfer dissociation tandem mass spectrometry, *Proc. Natl. Acad. Sci. U. S. A.*, **106**, pp. 995–1000.

75. Tani, K., and Ozawa, M. (1999). Investigation of chromatographic properties of titania. I. On retention behavior of hydroxyl and other substituent aliphatic carboxylic acids: Comparison with zirconia, *J. Liq. Chromatogr. Relat. Technol.*, **22**, pp. 843–856.

76. Thingholm, T. E., Jensen, O. N., and Larsen, M. R. (2009). Analytical strategies for phosphoproteomics, *Proteomics*, **9**, pp. 1451–1468.

77. Thompson, A. J., Hart, S. R., Franz, C., Barnouin, K., Ridley, A., and Cramer, R. (2003). Characterization of protein phosphorylation by mass spectrometry using immobilized metal ion affinity chromatography with on-resin beta-elimination and Michael addition, *Anal. Chem.*, **75**, pp. 3232–3243.

78. Torta, F., Usuelli, V., Malgaroli, A., and Bachi, A. (2008). Proteomic analysis of protein *S*-nitrosylation, *Proteomics*, **8**, pp. 4484–4494.

79. Tsai, C. F., Wang, Y. T., Chen, Y. R., Lai, C. Y., Lin, P. Y., Pan, K. T., Chen, J. Y., Khoo, K. H., and Chen, Y. J. (2008). Immobilized metal affinity chromatography revisited: pH/acid control toward high selectivity in phosphoproteomics, *J. Proteome Res.*, **7**, pp. 4058–4069.

80. Wang, Q., Zhang, Y., Yang, C., Xiong, H., Lin, Y., Yao, J., Li, H., Xie, L., Zhao, W., Yao, Y., Ning, Z. B., Zeng, R., Xiong, Y., Guan, K. L., Zhao, S., and Zhao, G. P. (2010). Acetylation of metabolic enzymes coordinates carbon source utilization and metabolic flux, *Science*, **327**, pp. 1004–1007.

81. Wang, Y., Liu, T., Wu, C., and Li, H. (2008). A strategy for direct identification of protein *S*-nitrosylation sites by quadrupole time-of-flight mass spectrometry, *J. Am. Soc. Mass Spectrom.*, **19**, pp. 1353–1360.

82. Wang, Y. T., Tsai, C. F., Hong, T. C., Tsou, C. C., Lin, P. Y., Pan, S. H., Hong, T. M., Yang, P. C., Sung, T. Y., Hsu, W. L., and Chen, Y. J. (2010). An informatics-assisted label-free quantitation strategy that depicts phosphoproteomic profiles in lung cancer cell invasion, *J. Proteome Res.*, **9**, pp. 5582–5597.

83. Welchman, R. L., Gordon, C., and Mayer, R. J. (2005). Ubiquitin and ubiquitin-like proteins as multifunctional signals, *Nat. Rev. Mol. Cell Biol.*, **6**, pp. 599–609.

84. Wolschin, F., Wienkoop, S., and Weckwerth, W. (2005). Enrichment of phosphorylated proteins and peptides from complex mixtures using metal oxide/hydroxide affinity chromatography (MOAC), *Proteomics*, **5**, pp. 4389–4397.

85. Yang, X. J. (2005). Multisite protein modification and intramolecular signaling, *Oncogene*, **24**, pp. 1653–1662.

86. Yang, X. J., and Seto, E. (2008). Lysine acetylation: codified crosstalk with other posttranslational modifications, *Mol. Cell*, **31**, pp. 449–461.

87. Yang, Y. Y., Ascano, J. M., and Hang, H. C. (2010). Bioorthogonal chemical reporters for monitoring protein acetylation, *J. Am. Chem. Soc.*, **132**, pp. 3640–3641.

88. Yu, B. J., Kim, J. A., Moon, J. H., Ryu, S. E., and Pan, J. G. (2008). The diversity of lysine-acetylated proteins in *Escherichia coli*, *J. Microbiol. Biotechnol.*, **18**, pp. 1529–1536.

89. Zaia, J. (2004). Mass spectrometry of oligosaccharides, *Mass Spectrom. Rev.*, **23**, pp. 161–227.

90. Zhang, H., Li, X. J., Martin, D. B., and Aebersold, R. (2003). Identification and quantification of N-linked glycoproteins using hydrazide chemistry, stable isotope labeling and mass spectrometry, *Nat. Biotechnol.*, **21**, pp. 660–666.

91. Zhang, J., Sprung, R., Pei, J., Tan, X., Kim, S., Zhu, H., Liu, C. F., Grishin, N. V., and Zhao, Y. (2009). Lysine acetylation is a highly abundant and evolutionarily conserved modification in *Escherichia coli*, *Mol. Cell. Proteomics*, **8**, pp. 215–225.

92. Zhao, S., Xu, W., Jiang, W., Yu, W., Lin, Y., Zhang, T., Yao, J., Zhou, L., Zeng, Y., Li, H., Li, Y., Shi, J., An, W., Hancock, S. M., He, F., Qin, L., Chin, J., Yang, P., Chen, X., Lei, Q., Xiong, Y., and Guan, K. L. (2010). Regulation of cellular metabolism by protein lysine acetylation, *Science*, **327**, pp. 1000–1004.

93. Zhao, Y., and Jensen, O. N. (2009). Modification-specific proteomics: strategies for characterization of post-translational modifications using enrichment techniques, *Proteomics*, **9**, pp. 4632–4641.

Chapter 5

Cardiovascular Proteomic Analysis

Toru Suzuki[a,b] **and Ryozo Nagai**[a]
[a]*Department of Cardiovascular Medicine, The University of Tokyo, 7-3-1 Hongo, Bunkyo-ku, Tokyo 113-8655, Japan*
[b]*Department of Ubiquitous Preventive Medicine, The University of Tokyo, 7-3-1 Hongo, Bunkyo-ku, Tokyo 113-8655, Japan*
torusuzu-tky@umin.ac.jp

In this chapter, we summarize our proteomic studies in the field of cardiovascular medicine. Our research focuses on understanding the role of proteins in cardiovascular disease with an objective of better understanding cardiovascular pathophysiology to lead to the development of new and better diagnostic and therapeutic methods. A particular focus has been placed on understanding intracellular pathogenic processes such as epigenetic regulation and extracellular processes such as identifying and quantitating extracellular proteins. We have used mass spectrometry as both an exploratory and a diagnostic technology for the detection of processes ranging from protein–protein interactions to post-translational modifications in cellular proteins as well as circulating biomarkers. Fractionation methods for proteins such as liquid-based two-dimensional methods as well as immuno-enrichment procedures have also been explored

Quantitative Proteome Analysis: Methods and Applications
Edited by Kazuhiro Imai and Sam F. Y. Li

ISBN 978-981-4316-51-4 (Hardcover), 978-981-4316-98-9 (eBook)
www.panstanford.com

to enable further facilitation of protein pre-separation. In addition to these functional proteomics studies, structural proteomic studies are also done with a hope to allow for pinpoint drug design and therapeutic intervention. Collectively, our proteomics studies are focused on understanding the functional role and potential therapeutically exploitable property of proteins in cardiovascular disease from both intracellular and extracellular aspects with both functional and structural proteomics approaches to allow for comprehensive analysis.

5.1 Introduction

Major interests in the life sciences have been rapidly transitioning to understanding proteins as the major field of research in the post-genomic era. The completion of the human genome project demonstrated the marginal increase in genes during evolution that diversity in humans is likely dictated by post-genomic regulation, namely proteins, [1, 2]. Aside from the recently addressed field of RNA regulation (e.g. micro-RNA, etc.), post-genomic regulation at the protein level is likely the major regulatory step. Splicing, processing and post-translational modifications in addition to the protein–protein interaction are but some of the notable regulatory pathways.

In contrast to genes/DNA, which are quite stable as would be necessary for conserved preservation of the genetic information, proteins, which are the products of genes, are rather unstable and have short half-lives being produced and then degraded to be present to function only when and where their need is dictated as reflective of their physiologically active properties. Not only are different proteins produced at times through splicing, but once they are produced they are often processed/cleaved and/or modified (e.g. phosphorylation, glycosylation), and act in concert with other proteins as complexes before they are degraded. Indeed, understanding the complex regulation of proteins under a given condition (e.g. temporospatial context) is of utmost importance in understanding their function.

While proteomic analysis in general concerns the comprehensive understanding of proteins and their functions, which often entails understanding their functions in a non-challenged or quiescent state, our interests are focused on the role of proteins as relevant in disease states and their regulation as manifested under disease conditions. Temporal changes in disease-associated factors are important in understanding pathophysiology. For instance, genetic disorders as represented by monogenic disorders often manifest phenotypes in early life. However, for acquired diseases such as oncogenic as well as lifestyle-oriented diseases, environmental factors play an important contributory role to onset of disease aside from their genetic underpinnings. Although the regulation of gene expression undoubtedly plays an important role, temporal changes in later/adult life is regulated in main at the level of proteins, and therefore studies at the protein level are ideal for studies focused on disease onset and progression. We therefore strive to understand the molecular changes which affect proteins under disease conditions, not only to understand their role in the pathophysiological basis of disease, but to further exploit their possible roles in diagnostic as well as therapeutic applications.

The following discussion on our use of proteomic approaches will be separated into categories of analysis on intracellular proteins and extracellular proteins.

5.2 Proteomic Analysis of Intracellular Proteins

For the proteomic analysis of intracellular proteins, we have focused on understanding intracellular pathogenic processes with a particular focus on understanding epigenetic regulation in cardiovascular disease [3, 4]. To briefly provide the background on epigenetic regulation, in eukaryotes, including mammals such as humans, the genomic DNA is packaged into chromatin, whose fundamental unit is the nucleosome in which DNA is twice wrapped around the histone octamer. This compaction of DNA is likely necessary to efficiently package the vast amount of genomic DNA as found in the eukaryotic cell. This adds an additional step necessary in activation of transcription as DNA is wound in the chromatin state under basal conditions. Only after the chromatin structure is relieved can transcription factors, including both regulatory and general factors, access the promoter to regulate transcription and thus gene expression.

Epigenetic regulation impacts a multitude of physiological as well as pathophysiological reactions ranging from cancer to cardiovascular disease. It had been poorly addressed in the field of cardiovascular disease likely due to more important focus on physiology as well as membrane-oriented cellular biology. However, pioneering work in the skeletal muscle, which is the prototype for cardiac as well as smooth muscle, which are relevant in the cardiovasculature, suggested from early on that epigenetic regulation may be important in phenotypic regulation. Studies using the DNA methylation inhibitor, 5-azacytidine, showed that the regulation of methylation state can induce myocyte differentiation [5]. This work led to the discovery of the transcription factor, MyoD, a master regulator transcription factor of myogenic differentiation [6]. Recent studies have shown that epigenetic regulation, namely in the form of regulators of the post-transcriptional modification, acetylation, play a role in regulating cardiac growth and re-modelling (e.g. cardiac hypertrophy and failure). The studies from the Olson lab have been instrumental in establishing that acetylation/deacetylation, in particular through the functions of deacetylases (HDACs, histone deacetylases), play a pivotal role in cardiovascular phenotypic modulation and disease [7, 8].

We have also focused in parallel on the role of epigenetic regulation of cardiovascular disease with a focus on understanding the role of protein–protein interaction and post-translational modification, namely acetylation. For this, we have used proteomic approaches to understand the underlying mechanisms of functional regulation of the Sp- and Kruppel-like factors of zinc finger transcription factors, which play an important role in cardiovascular regulation [3, 9]. As protein–protein interaction and post-translational modifications are detectable by proteomic approaches, as a result, we were able to define novel mechanisms of epigenetic regulation of transcription and cardiovascular disease through our studies, which will be discussed in detail below.

Zinc finger factors, especially those which contain a cysteine-histidine (Cys2-His2) zinc-finger motif, emerged through evolution and markedly increased in eukaryotes [10]. Although this common motif is widely found in cellular factors ranging from enzymes to transcription factors, often the motif is used as a DNA-binding domain in transcription factors. As these factors preferentially evolved in eukaryotes, they are thought to play functional roles in developmental and differentiation processes among others. The specificity protein- (Sp-) and Kruppel-like factors (KLFs) are a family of these factors which is centred on Sp1 (specificity protein-1), which was one of the first regulatory transcription factors identified as a factor which stimulates the simian virus 40 (SV40) early promoter by binding to its GC-rich promoter sites [11]. The subsequent categorization of the genome showed that Sp1 and its related factors show sequence similarity centred on the zinc-finger region with other zinc finger transcription factors related to the Kruppel gene as identified in Drosophila as being a gap segmentation gene [12–19]. We now understand that there are approximately 20 of these factors in mammals, and through gene mutation studies that these factors often have individual biological functions. For instance, KLF1 (aka erythroid KLF, EKLF) is required for erythrocyte development. Interestingly, studies on zebrafish have shown that the KLFs have evolved to function mainly in the blood and circulatory organs [20]. Thus, the KLFs are an excellent target for our work, which centres often on the cardiovascular role of proteins.

We set out to understand the role of transcription factors in chromatin regulation with a focus on the Sp/KLF factors in cardiovascular pathophysiology. Recent studies have shown chromatin transcription is regulated by three classes of factors which include (1) modification enzymes such as methylases and acetylases, (2) ATP-independent nucleosome assembly factors which are also called histone chaperones and (3) ATP-dependent nucleosome assembly factors. These factors through protein–protein interactions and post-translational modifications affect their target protein (e.g. histones, transcription factor) to regulate transcription.

Using the zinc finger DNA-binding domain region, which is often an interface for protein–protein interactions, of Sp1 and KLF5, the latter being a Kruppel-like factor important in mediating cardiovascular re-modelling in response to external stress, we affinity purified proteins which interact with these peptides from nuclear extract and identified them by use of matrix-assisted laser desorption/ionization time-of-flight mass spectrometry (MALDI TOF-MS) with subsequent peptide mass fingerprinting/post-source decay (PMF/PSD) analysis. We identified an ATP-independent nucleosome assembly factor, template activating factor-I (TAF-I/SET) to interact with both Sp1 and KLF5 [21, 22]. TAF-I/SET inhibited the DNA-binding and promoter activation activities of Sp1 and KLF5 and thus acts as a transcriptional repressor. We had previously shown through a focused approach that Sp/KLF factors are differently acetylated. We found that the coactivator/acetylase p300 interacts and acetylates KLF5 to coactivate transcription, and interestingly that TAF-I/SET could mask KLF5 from being acetylated by p300. These showed that TAF-I/SET could inhibit acetylation of transcription factor. We also showed the regulation of these factors under pathophysiological conditions. Under phorbol ester stimulation (as an agonist of pathophysiological stimulation), KLF5 was upregulated as was its downstream gene, platelet-derived growth factor A-chain (PDGF-A), but interestingly the repressor, TAF-I/SET, was downregulated under these conditions, thus likely allowing for stimulation of transcription. As p300 has been shown to be induced by phorbol ester, it is likely that coordinated induction/repression of cofactor with transcription is important for transcriptional regulation. We further showed that protein–protein interaction (e.g. p300, TAF-I/SET) is coupled with post-translational modification (e.g. acetylation) and thus showed a novel mechanism of transcriptional regulation as defined by proteomic approaches. We further went on to show that the deacetylase, HDAC1, competes with p300 for interaction with KLF5, which showed that deacetylase regulates transcription at levels other than catalytic regulation [23]. More recently, we have identified a novel ATP-independent nucleosome assembly factor, ANP32B, to interact with KLF5 and to repress transcription at the nucleosome level [24]. Through our studies and those of others, we now know that the Sp/KLF factors, aside from histones, are the only family of factors which associate with all three classes of chromatin re-modelling factors, with further regulation among these classes of chromatin re-modelling factors [4]. Thus, proteomic approaches, which often excel in identifying protein–protein interactions as well as post-translational modifications, played an important role in identifying a novel pathophysiological transcriptional regulatory mechanism.

The DNA damage response pathway is also another aspect of epigenetic regulation that we have been pursuing. Initial studies to identify KLF5-interactors showed that the DNA damage sensor protein, poly (ADP-ribose) polymerase (PARP), interacts with KLF5 to regulate the anti-apoptotic functions of KLF5 in cardiovascular pathology [25]. As this finding that the DNA damage response pathway function in the cardiovasculature was somewhat surprising given that this pathway is known to play a major role in the checkpoint for cancer, we further investigated the contributory role of this pathway by investigating the role of the ataxia telangiectasia mutated (ATM) protein, which is the central molecule in this pathway. We recently showed that ATM is involved in regulating vascular senescence and aging [26]. Thus, the DNA damage response albeit initially unexpected plays a contributory role to cardiovascular pathophysiology

We have further carried out crystallographic studies to understand the biophysical mechanisms of the protein–protein interaction and post-translational modifications as stated with the ultimate goal to develop pinpoint drugs/compounds to specifically regulate chromatin transcription in the context of pathophysiological regulation. We have already solved the crystal structure of TAF-I/SET [27] and that of ANP32B [28]. Co-crystal structure analysis should allow for designing specific compounds. Thus, our functional and structural proteomic analysis is a model approach for comprehensive understanding of the role of proteins in cardiovascular pathophysiology.

In summary, proteomic approaches have been instrumental in the discovery of new pathophysiological mechanisms of cardiovascular disease involving protein–protein interaction and post-translational modification (acetylation). Given that our studies were done in a timely manner as concurrent with studies by others to study the role of epigenetic regulation with a focus on acetylation/deacetylation in the cardiovasculature, the field of cardiovascular research has much advanced in its comprehensive understanding of involved proteins in recent years. Importantly, in contrast to the work of others which has mainly focused on use of animal models, our proteomic studies have allowed for understanding the underlying pathogenic mechanisms at the protein level which complement these works done using different approaches. These mechanistic insights will likely lead to new approaches for therapeutic intervention. Cardiovascular disease now sets a benchmark for understanding epigenetic regulation in disease and will likely impact other fields (e.g. metabolic disease, cancer) which are diseases with similar acquired changes in later life. One concern which remains for proteomic analysis is that the abundant protein is the first to be identified, which may not necessarily physiologically relevant. More emphasis on combined functional assays with MS-based identification will be needed. We were fortunate that our selection of protein interaction domain and cell conditions resulted in favourable results. However, it is important to note that much trial and error was needed in establishing conditions which benefited much from our experience and training in classical cold-room techniques using chromatographic columns to separate proteins. Essentially, the MS-based techniques were much a technological advancement allowing for detection with less amounts of protein, but the questions and experimental logic for procedure used was much consistent with those of the past. Knowledge and experience in protein purification is thus likely a prerequisite for effective use of MS-based techniques. Further studies will be aimed at better understanding the temporo-spatial regulation of protein–protein interaction and post-translational modification in the combinatorial context (e.g. complex) with further understanding their roles *in situ* using imaging MS techniques which are being developed at present. The eventual development of an interaction and regulatory map of proteins in disease states, not only in cardiovascular medicine but in all medical fields, will surely allow for the better understanding of the pathophysiology as well as allow for targeted therapeutic intervention.

5.3 Proteomic Analysis of Extracellular Proteins

For extracellular proteins, we have focused on discovering new proteins and their functional roles for exploitation in diagnostic purposes (e.g. biomarkers) as well as for therapeutic purposes (e.g. bioactive molecules). Since the pioneering work of predecessors of the lab in which cardiac myosin light chain was purified and then applied for the diagnosis of myocardial infarction [29], clinical application of protein chemistry has been a longstanding research topic. The authors have also recently shown that vascular smooth muscle proteins can be used in the diagnosis of aortic dissection [30–33], which is the first diagnostic application of a blood test for this disease, as well as various oxidized LDL assays in the diagnosis of coronary artery disease [34–36], in addition to characterization of other cardiovascular diagnostic markers such as the natriuretic peptides among others (e.g. interleukin-6 in coronary artery disease) [37–39].

Thus, with this longstanding background in diagnostic applications of protein-based markers and assays, the authors were keen to introduce MS technology to the lab early on. We first used MALDI-TOF MS to identify post-translational modifications (e.g. acetylation) in the above-mentioned intracellular proteomic studies and also aimed at using this technology to identify new pathophysiological proteins in the cardiovasculature.

There are two critical determinants of using MS-technology for the detection of proteins. First is the detector as exemplified by MALDI-based MS detection as well as electrospray ionization-type detection (ESI). Importantly, MS technology allows for the detection of minute amounts of protein as would not have been possible by classical techniques. The other important determinant is fractionation or pre-MS separation procedures. This is the most important step for any protein identification or detection procedure, either being a classical technique or recent one. Finally, developing a workflow which best suits the specific needs of the lab must be developed based on these steps. For instance, if working with blood samples, a further workflow involving sample preparation prior to the separation procedure must be optimized.

Our studies have focused on addressing the possibility of using MS-based technologies not only as a research tool but also as a diagnostic tool for clinical medicine with a particular focus on blood-based techniques. On the former point of the detector, while we also use ESI-based techniques as a research tool, we have focused in main on using MALDI-based detection. At present, five devices are used in the lab, which include a PerSeptive Biosystems Voyager DE-STR, Ciphergen ProteinChip PBSII reader, and Shimadzu Biotech AXIMA QIT, CFR and Confidence instruments. Our findings and opinions are based primarily by use of such equipment, and workflow as optimized for such. Clearly, the more sophisticated and sensitive instrumentation provides more information on the sample, but questions still remain on quantitative assessment, reproducibility (between-run) and standardization as would be necessary for diagnostic use. Further, the specifications as would be needed for diagnostic detection have yet to be determined. As our knowledge and experience with this technology advances and specific diagnostic content and protocols become established, then the specifications as necessary for diagnostic use will become apparent. Only then will a MS device specific for clinical diagnostic use become available, and most likely be refined for this purpose. One further important issue for clinical diagnostic use would be that the pricing of MS devices decrease to a reasonable level to allow wide use as well as ease-of-use, but this will likely be dictated by the necessary specifications needed.

The issue of protein fractionation has remained the most challenging aspect. In blood, approximately 20 proteins, including albumin, IgG and haptoglobin among others, comprise the majority of the proteins available. At present, methods are available to facilitate specific removal of these abundant proteins from blood (e.g. Agilent, Beckman Coulter), but at times the protein of interest may be attached to these abundant proteins for which clearance may not necessarily be wise. Ion exchange and other protein purification procedures can be applied at micro-scale levels to separate proteins as well. One technology which warrants mention is the surface-enhanced laser desorption ionization (SELDI) MS technology (Ciphergen/Bio-Rad), which is characterized by chemically modified surfaces similar to chromatography (e.g. ion-exchange, metal-affinity, etc.), but importantly manipulated on the MS plate thus allowing for both separation of protein as well as subsequent detection. This seemed to be an advantageous solution to a single workflow for on-chip fractionation and high-throughput analysis, although a verdict has yet to be reached. Numerous diagnostic applications of this technology are presently under investigation mainly in the field of cancer (e.g. ovarian cancer, prostate cancer) [40, 41], as are the studies which address its reproducibility (e.g. different centres) [42] as would be necessary for a diagnostic platform. Although an approved diagnostic use has yet to become available, studies have shown that MS-based diagnosis may provide early and sensitive detection of ovarian cancer as compared to the classical tumour marker, CA125, as an example [40].

Two-dimension electrophoresis (2D-PAGE) and immuno-based separation procedures remain major areas of research. 2D-PAGE has been classically used for protein separation as well as differential analysis, and presently labelling methods as well as computer-assisted identification procedures have much facilitated this technology. It is important to note that MS technology is most optimized for the detection of peptides and small molecules (e.g. <5000 daltons) and when proteins and larger molecules are of interest, 2D-PAGE separation still remains a viable procedure despite its low-throughput. How to improve upon 2D separation procedures therefore remains an important topic. One recent technology which warrants mention is the PF 2D system (Beckman Coulter), which allows separation of proteins by two-dimensional liquid chromatography system based on separation by chromatofocussing with subsequent reversed-phase chromatography. We performed differential proteomic analysis using an animal model of diabetes mellitus and associated metabolic disorders (Otsuka Long-Evans Tokushima Fatty rat) using this system. Differentially expressed proteins in serum were identified with MALDI-TOF mass spectrometry including apolipoproteins and alpha2-HS-glycoprotein [43]. This was the first application of this approach to differential serum proteomics. While this technique still is in early stages and has poor throughput, it shows the potential of non-gel-based 2D techniques in identifying disease-associated proteins. We further used this technology to identify a new regulatory pathway of transforming growth factor-beta in vascular disease [44].

We also helped develop a micro-scale 2D device for clinical use as initially described [45]. As this technology could allow liquid-based separation according to pI-values with high throughput and as the chips are disposable, there are high expectations for clinical applications.

Antibody-based immuno-separation is also an attractive procedure once the protein of interest is identified. Optimal procedures and platforms (e.g. beads, plates) for use of immuno-based separation will also need to be addressed. Importantly, as protein fragments may be associated with disease as well as their modifications, immuno-based isolation will likely be pivotal for clinical application. For instance, fragmented and modified forms of the cardiac troponin protein are known to be associated with cardiac disease [46, 47].

Whether MS instrumentation will find its way into mainstream diagnosis remains a question to be answered, but it is already used for the diagnosis of inborn metabolic disorders and familial amyloidosis among others. For this technology to become a viable addition to diagnostic testing will require that it shows that it is cost-effective and shows important information which cannot be detected by any other technique. For instance, protein fragmentation and post-translational modification in disease are surely attractive targets in which detection by MS would excel, and for this to be a viable test would likely require that a panel of antibodies (e.g. protein chip) become available to allow for effective detection (e.g. cost, time).

Quantization, reproducibility and portability are also issues which remain to be addressed.

5.4 Conclusions: Implications and Lessons Learned

In all, this review has summarized our approaches to proteomic medicine, and how we have used it in our lab to better understand cardiovascular pathophysiology with a focus on epigenetic regulation as well as biomarker discovery with further interest in clinical application of the technology. Although clinical application or exploitation of MS technology is still in young stages, further advancements will surely clarify the use of this biotechnology as a medical tool.

In a general context, the cardiovascular proteome is a current topic of interest and in the United States the National Heart, Lung, and Blood Institute (NHLBI) has ongoing a clinical proteomics working group that is charged with identifying opportunities and challenges in clinical proteomics and using these as a basis for recommendations aimed at directly improving patient care [48]. The Human Proteome Organisation (HUPO) plasma proteome project also had a cardiovascular sub-study which catalogued plasma proteins to be used as a resource [49]. HUPO has just announced the start of the Human Proteome Initiative at the annual meeting (September 2010, Sydney, Australia) and aims to elucidate the human proteome during the next decade. Other studies have been ongoing on a global level to address the technologies and their results as would be relevant to cardiovascular disease. Although most studies have resulted in cataloguing differential protein profiles in cells and disease states [50, 51], this is an important initial step in determining the potential of the technology and cross-referencing of results. Given the various technologies and instrumentation available at present, these studies should be of value in determining which method meets the needs for specific aims. As we learn through our experiences in parallel with the advancements in technologies and instrumentation, this field will surely rapidly evolve within the coming years. Eventually, we aim to have a comprehensive understanding of protein dynamics (e.g. interaction, modification, degradation) in disease at both intra- and extracellular levels with ultimate goals of better understanding pathophysiological mechanisms, which will lead to development of new diagnostic and therapeutic techniques.

Acknowledgements

This study was supported by grants from the Ministry of Education, Culture, Sports, Science, and Technology; New Energy and Industrial Technology Development Organization; and the Ministry of Health, Labour, and Welfare in addition to grants from the following foundations: Applied Enzymology, Takeda Science and Japan Heart.

References

1. Venter, J. C., and Adams, M. D., *et al.* (2001). The sequence of the human genome, *Science.* **291**, p. 1304.
2. McPherson, J. D., and Marra, M., *et al.* (2001). A physical map of the human genome, *Nature.* **409**, pp. 934–941.
3. Suzuki, T., and Horikoshi, M., *et al.* (2003). Regulation of the Sp/KLF-family of transcription factors: focus on post-transcriptional modification and protein–protein interaction in the context of chromatin, *Gene. Ther. Mol. Biol.* **7**, pp. 91–98.
4. Suzuki, T., and Matsumura, T., *et al.* (2005). Transcriptional regulation at the chromatin level in the cardiovasculature through protein–protein interactions and chemical modifications, *Trends. Cardiovasc. Med.* **15**, pp. 125–129.
5. Davis, R. L., and Weintraub, *et al.* (1987). Expression of a single transfected cDNA converts fibroblasts to myoblasts, *Cell.* **51**, pp. 987–1000.
6. Tapscott, S. J., and Davis, R. L., *et al.* (1988). MyoD1: a nuclear phosphoprotein requiring a Myc homology region to convert fibroblasts to myoblasts, *Science.* **242**, pp. 405–411.
7. Olson, E. N. (2006). Gene regulatory networks in the evolution and development of the heart, *Science.* **313**, pp. 1922–1927.
8. Backs, J., and Olson, E. N. (2006). Control of cardiac growth by histone acetylation/deacetylation, *Circ. Res.* **6**, pp. 15–24.
9. Suzuki, T., and Aizawa, K., *et al.* (2005). Vascular implications of the Krüppel-like family of transcription factors, *Arterioscler. Thromb. Vasc. Biol.* **25**, pp. 1135–1141.

10. Tupler, R., and Perini G., *et al.* (2001). Expressing the human genome, *Nature.* **409**, pp. 832–833.
11. Dynan, W. S., and Tjian, R. (1983). The promoter-specific transcription factor Sp1 binds to upstream sequences in the SV40 early promoter, *Cell.* **35**, pp. 79–87.
12. Philipsen, S., and Suske, G. (1999). A tale of three fingers: the family of mammalian Sp/XKLF transcription factors, *Nucleic. Acids. Res.* **27**, pp. 2991–3000.
13. Turner, J., and Crossley, M. (1999). Mammalian Krüppel-like transcription factors: more than just a pretty finger, *Trends. Biochem. Sci.* **24**, pp. 236–240.
14. Dang, D.T., and Pevsner, J., *et al.* (2000). The biology of the mammalian Krüppel-like family of transcription factors, *Int. J. Biochem. Cell. Biol.* **32**, pp. 1103–1121.
15. Bieker, J. J. (2001). Krüppel-like factors: three fingers in many pies, *J. Biol. Chem.* **276**, pp. 34355–34358.
16. Black, A. R., and Black, J. D., *et al.* (2001). Sp1 and krüppel-like factor family of transcription factors in cell growth regulation and cancer, *J. Cell. Physiol.* **188**, pp. 143–160.
17. Bouwman, P., and Philipsen, S. (2002). Regulation of the activity of Sp1-related transcription factors, *Mol. Cell. Endocrinol.* **195**, pp. 27–38.
18. Kaczynski, J., and Cook, T. (2003). Sp1- and Krüppel-like transcription factors, *Genome. Biol.* **4**, p. 206.
19. Feinberg, M.W., and Lin, Z., *et al.* (2004). An emerging role for Krüppel-like factors in vascular biology, *Trends. Cardiovasc. Med.* **14**, pp. 241–246.
20. Oates, A. C., and Pratt, S. J., *et al.* (2001). The zebrafish klf gene family, *Blood.* **98**, pp. 1792–1801.
21. Suzuki, T., and Muto, S., *et al.* (2003). Functional interaction of the DNA-binding transcription factor Sp1 through its DNA-binding domain with the histone chaperone TAF-I, *J. Biol. Chem.* **278**, pp. 28758–28764.
22. Miyamoto, S., and Suzuki, T., *et al.* (2003). Positive and negative regulation of the cardiovascular transcription factor KLF5 by p300 and the oncogenic regulator SET through interaction and acetylation on the DNA-binding domain, *Mol. Cell. Biol.* **23**, pp. 8528–8541.
23. Matsumura, T., and Suzuki, T., *et al.* (2005). The deacetylase HDAC1 negatively regulates the cardiovascular transcription factor Krüppel-like factor 5 through direct interaction, *J. Biol. Chem.* **280**, pp. 12123–12129.

24. Munemasa, Y., and Suzuki, T., *et al.* (2008). Promoter region-specific histone incorporation by the novel histone chaperone ANP32B and DNA-binding factor KLF5, *Mol. Cell. Biol.* **28**, pp. 1171–1181.

25. Suzuki, T., and Nishi, T., *et al.* (2007). Functional interaction between the transcription factor Krüppel-like factor 5 and poly(ADP-ribose) polymerase-1 in cardiovascular apoptosis, *J. Biol. Chem.* **282**, pp. 9895–9901.

26. Zhan, H., and Suzuki, T., *et al.* (2010). Ataxia telangiectasia mutated (ATM)-mediated DNA damage response in oxidative stress-induced vascular endothelial cell senescence, *J. Biol. Chem.* **285**, pp. 29662–29670.

27. Muto, S., and Senda, M, *et al.* (2004). Purification, crystallization and preliminary X-ray diffraction analysis of human oncoprotein SET/TAF-1 β, *Acta. Crystallogr. Sect. D- Biol. Crystallogr.* **60**, pp. 712–714.

28. Tochio, N., and Umehara, T., *et al.* (2010). Solution structure of histone chaperone ANP32B: interaction with core histones H3-H4 through its acidic concave domain, *J. Mol. Biol.* **401**, pp. 97–114.

29. Nagai, R., and Ueda, S., *et al.* (1979). Radioimmunoassay of cardiac myosin light chain II in the serum following experimental myocardial infarction, *Biochem. Biophys. Res. Commun.* **86**, pp. 683–688.

30. Katoh, H., and Suzuki, T., *et al.* (1995). Diagnosis of aortic dissection by immunoassay for circulating smooth musclemyosin, *Lancet.* **345**, pp. 191–192.

31. Suzuki, T., and Katoh, H., *et al.* (1996). Novel biochemical diagnostic method for aortic dissection. Results of a prospective study using an immunoassay of smooth muscle myosin heavy chain, *Circulation.* **93**, pp. 1244–1249.

32. Suzuki, T., and Katoh, H., *et al.* (1997). Biochemical diagnosis of aortic dissection by raised concentrations of creatine kinase BB-isozyme, *Lancet.* **350**, pp. 784–785.

33. Suzuki, T., and Katoh, H., *et al.* (2000). Diagnostic implications of elevated levels of smooth-muscle myosin heavy-chain protein in acute aortic dissection. The smooth muscle myosin heavy chain study, *Ann. Intern. Med.* **133**, pp. 537–541.

34. Toshima, S., and Hasegawa, A., *et al.* (2000). Circulating oxidized low density lipoprotein levels. A biochemical risk marker for coronary heart disease, *Arterioscler. Thromb. Vasc. Biol.* **20**, pp. 2243–2247.

35. Suzuki, T., and Kohno, H., *et al.* (2002). Clinical Trial of Oxidized LDL (CTOL) investigators. Diagnostic implications of circulating oxidized

low density lipoprotein levels as a biochemical risk marker of coronary artery disease, *Clin. Biochem.* **35**, pp. 347–353.

36. Amaki, T., and Suzuki, T., *et al.* (2004). Circulating malondialdehyde modified LDL is a biochemical risk marker for coronary artery disease, *Heart.* **90**, pp. 1211–1213.

37. Suzuki, T., and Yamaoki, K., *et al.* (2000). Screening for cardiac dysfunction in asymptomatic patients by measuring B-type natriuretic peptide levels, *Jpn. Heart. J.* **41**, pp. 205–214.

38. Suzuki, T., and Yamazaki, T., *et al.* (2001). The role of the natriuretic peptides in the cardiovascular system, *Cardiovasc. Res.* **51**, pp. 489–494.

39. Suzuki, T., and Ishiwata, S., *et al.* (2000). Raised interleukin 6 concentrations as a predictor of postangioplasty restenosis, *Heart.* **83**, p. 578.

40. Zhang, Z., and Bast Jr, R. C., *et al.* (2004). Three biomarkers identified from serum proteomic analysis for the detection of early stage ovarian cancer, *Cancer. Res.* **64**, pp. 5882–5890.

41. Adam, B. L., and Qu, Y., *et al.* (2002). Serum protein fingerprinting coupled with a pattern-matching algorithm distinguishes prostate cancer from benign prostate hyperplasia and healthy men, *Cancer. Res.* **62**, pp. 3609–3614.

42. Semmes, Z., and Feng, B. L., *et al.* (2005). Evaluation of serum protein profiling by surface-enhanced laser desorption/ionization time-of-flight mass spectrometry for the detection of prostate cancer: I. Assessment of platform reproducibility, *Clin. Chem.* **51**, pp. 102–112.

43. Matsumura, T. and Suzuki, T., *et al.* (2006). Differential serum proteomic analysis in a model of metabolic disease, *Biochem. Biophys. Res. Commun.* **351**, pp. 965–971.

44. Matsumura, T., and Suzuki, T., *et al.* (2009). Regulation of transforming growth factor-beta-dependent cyclooxygenase-2 expression in fibroblasts, *J. Biol. Chem.* **284**, pp. 35861–35871.

45. Fujita, M., and Hattori, W., *et al.* (2006). High-throughput and high-resolution two dimensional mapping of pI and m/z using a microchip in a matrix-assisted laser desorption/ionization time-of-flight mass spectrometer, *J. Chromatogr. A.* **1111**, pp. 200–205.

46. Labugger, R., and Organ, L., *et al.* (2000). Extensive troponin I and T modification detected in serum from patients with acute myocardial infarction, *Circulation.* **102**, pp. 1221–1226.

47. McDonough, J. L., and Van Eyk, J. E. (2004). Developing the next generation of cardiac markers: disease-induced modifications of troponin I, *Prog. Cardiovasc. Dis.* **47**, pp. 207–216.

48. Granger, C. B., and Van Eyk, J. E., *et al.* (2004). National heart, lung, and blood institute clinical proteomics working group report, *Circulation.* **109**, pp. 1697–1703.

49. Berhane, B. T., and Zong, C., *et al.* (2005). Cardiovascular-related proteins identified in human plasma by the HUPO Plasma Proteome Project pilot phase, *Proteomics.* **5**, pp. 3520–3530.

50. Westbrook, J. A., and Wheeler, J. X., *et al.* (2006). The human heart proteome: Two-dimensional maps using narrow-range immobilised pH gradients, *Electrophoresis.* **27**, pp. 1547–1555.

51. Mayr, U., and Mayr, M., *et al.* (2005). Proteomic dataset of mouse aortic smooth muscle cells, *Proteomics.* **5**, pp. 546–557.

Chapter 6

The Proteome in Neurodegenerative Diseases

Teruyuki Tsuji,[a] Masahiro Aoki,[b] and Shun Shimohama[c]
[a]*Department of Neurology, Graduate School of Medicine, Kyoto University, 54 Shogoin-Kawaharacho, Sakyoku, Kyoto 606-8507, Japan*
[b]*Kazusa Laboratory, Protein Express Co., Ltd, Kisarazu, Chiba 292-0818, Japan*
[c]*Department of Neurology, School of Medicine, Sapporo Medical University, South-1, West-16, Chuo-ku, Sapporo 060-8543, Japan*
shimoha@sapmed.ac.jp

6.1 Introduction

There have been numerous achievements over the last few decades in the study of the molecular events associated with physiological and pathological processes. These have been made using genomic, transcriptomic, and proteomic approaches. It is now possible to examine the brain, the most complex organ, using proteomic approaches to obtain comprehensive protein profiles. Although there are several issues to address, including the lack of standardized sample preparation techniques and proper bioinformatics tools, proteomics is still progressing, and it has allowed researchers to compare

Quantitative Proteome Analysis: Methods and Applications
Edited by Kazuhiro Imai and Sam F. Y. Li

ISBN 978-981-4316-51-4 (Hardcover), 978-981-4316-98-9 (eBook)
www.panstanford.com

protein expression patterns with the genome and the transcriptome in various brain diseases [1]. In particular, proteomic studies of the brain—neuroproteomics—has helped us to understand the cellular, molecular, and functional complexity of the brain, and it can address the dynamic organization of protein networks and macromolecular structures [2]. The field of neuroproteomics faces special challenges due to the complex cellular and sub-cellular architecture of the central nervous system (CNS) [3], and the quest for biomarkers in nervous system diseases is a crucial task for researchers [4].

With aging, the prevalence of neurodegenerative diseases, such as Alzheimer's disease (AD), Parkinson's disease (PD), and even amyotrophic lateral sclerosis (ALS), rises sharply, and the burden related to these diseases will increasingly challenge public health systems. In spite of the progress made in understanding the molecular cascades affected in CNS diseases, the pathogenesis of neurodegeneration remains largely unknown.

In AD, pathological changes in the brain may occur as early as 20–30 years prior to the onset of clinical symptoms [5, 6]. Currently available drugs are able to delay the symptoms and progression of the disease and to improve the quality of life of both patients and their caregivers [7]. Treatment should be initiated as early as possible to provide optimal benefit, but unfortunately, the diagnosis of early AD is difficult. Therefore, diagnostic biomarkers, which facilitate the early diagnosis, monitoring, and prognosis of a variety of chronic neurodegenerative processes, have long been awaited.

Classical approaches for biomarker discovery are based on biochemical research on potential pathological processes, and on the selection of possible candidate markers, as was the case for β-amyloid and tau in AD, α-synuclein in PD, and TDP43 in ALS. Besides these targeted approaches using established pathological hallmarks, technical advances over the last decade have revealed the potential of comprehensive multiparametric approaches for biomarker discovery. While the data obtained from genomics may explain why certain individuals have an increased risk of acquiring a certain disease, there is no blueprint describing how the disease process may be triggered later on in life. Environmental (non-gene-mediated) mechanisms are thought to play critical roles in the onset of disease.

Considerable advances in recent years in human proteomic analysis have made it possible to overcome several obstacles hindering clinical application. In clinical practice, of utmost importance is the source of material to be used. Biomarker sets will help eliminate individual marker variability to achieve diagnostic precision. Furthermore, given the overlapping neuropathological profiles of parkinsonian dementia, ALS-associated dementia and multisystem atrophy, a multivariate proteomic marker approach should allow disease-to-disease comparison, making it possible to find a common proteomic profile among AD, PD, and ALS, as well as to determine the proteomic features that are unique to the individual diseases, ultimately permitting a critical comparison of these neurodegenerative diseases.

The clinical application of proteomic analysis faces a challenge in the statistical evaluation of datasets in which combinations of single markers are assessed using specific multivariate biomarker models. However, due to advances in technology, the variables that can be analyzed have increased sharply, providing enormous potential benefits. When biomarkers begin to be analyzed with sufficient accuracy, a single analysis might be sufficient to properly assess pathophysiological status.

Neurodegenerative diseases are characterized by the symmetrical loss of neurons in motor, sensory, and/or cognitive structures. AD, PD, and ALS, among other neurodegenerative disorders, are heterogeneous diseases with insidious and overlapping symptoms. The clinical diagnosis is usually delayed until the advanced stage of the disease. For AD and PD, a definitive neuropathological diagnosis can only be made by post-mortem examination of the brain [8]. As modern medicine can slow disease progression and alleviate clinical symptoms, it is of utmost importance to find biomarkers that can distinguish between the neurodegenerative diseases to permit early treatment.

In the following section, after a concise overview of the technical aspects of proteomics, we will review proteomic research on aging, the different sources of samples used for clinically oriented analysis, and the main technological approaches taken. Additionally, we will briefly describe our attempt to overcome the different challenges and the approaches taken for proteomic study of body fluids.

6.2 Biological Samples

For clinical proteomics, the selection of sample material is quite important. In principle, tissues and body fluids, as well as cultured cells, can be analyzed. For tissue samples that are closely associated with pathological changes, improvements in the field of laser capture microdissection allow sampling of homogenous tissue cell populations, but tissue harvesting requires invasive procedures. Consequently, in neuroproteomics, post-mortem tissue samples are usually used. In contrast, access to many body fluids requires minimal (e.g., blood and CSF) or no invasiveness (e.g., urine).

In general, proteins in biological samples can be degraded by proteolytic activity. Blood serum contains high proteolytic activity that is activated immediately on clotting, resulting in rapid degradation and the generation of proteolytic breakdown products. Therefore, the Human Proteome Consortium (HUPO) recommends analyzing blood in the form of plasma rather than serum and has established a standardized sample collection protocol [9]. However, even in blood plasma, considerable residual protease activity can be observed, which can interfere with proteomic analysis. In contrast, urinary proteins have been demonstrated to be stable enough for reliable proteomic analysis [10, 11]. Fiedler *et al.* [12] demonstrated the various effects of endogenous and exogenous variables on urine peptide patterns, such as urine sampling conditions, storage, freeze-thaw cycles, and blood/bacterial contamination. To minimize these effects, optimal protocols for urine sampling have been designed [13].

While unprocessed biological samples can be considered ideal for proteomic profiling, by avoiding artifacts due to the multiple steps of sample preparation, the presence of interfering compounds such as protein aggregates, albumin, lipids, carbohydrates, ions is a technical issue that remains to be overcome in sample preparation. Consequently, we should choose the sample preparation procedure carefully, considering that 1 additional step during preparation can add additional artifacts to the complex system. Furthermore, reproducibility of sample preparation is one of the most important considerations and cannot be overemphasized.

6.3 Sample Separation

Over the last decade, five different proteomic technology platforms have been commonly used for protein separation: two-dimensional gel electrophoresis (2DE), surface-enhanced laser desorption/ionization (SELDI), LC-MS, CE coupled to MS (CE-MS), and protein arrays.

In 2DE, proteins are separated according to their isoelectric point and molecular weight. Protein identification is routinely performed on proteolytic in-gel digests, followed by peptide extraction from the gel and MS analysis [14]. To improve gel-to-gel protein spot matching accuracy and to increase the reliability of protein quantification, two-dimensional difference gel electrophoresis (2D-DIGE) was developed. The method uses two samples differentially labeled with fluorescent dyes (e.g., Cy3 and Cy5). Subsequently, the two samples are separated simultaneously within the same gel. In addition to satisfactory comparison of two samples [15], the comparison of several different experiments become possible using another labeled sample as a calibrator. Methodologically, 2DE is rather time consuming, due to a certain extent to the lack of automation, and comparability between laboratories has been difficult to achieve [16]. However, 2DE still appears to be the method of choice for comparative analysis of abundant proteins.

SELDI was devised to overcome some of the limitations of 2DE-MS and has been applied in a number of clinical investigations of the proteome. Numerous matrices are used, such as hydrophilic materials, reverse-phase materials, and affinity reagents. Low reproducibility of binding, the inability to detect low abundance peptides and proteins, and the inability to directly identify discriminatory peaks are major issues to be overcome. With the progress in high-resolution mass spectrometry and improvements in sample preparation and pre-fractionation strategies, efforts are on the way to enhance reproducibility [17, 18].

LC-MS provides high-resolution separation with a large capacity for analytes that can be loaded onto an LC column. Approaches like multidimensional protein identification technology [19] and 2D liquid phase fractionation [20] provide a great amount of data. However, LC-MS tends to be time consuming, which limits its application in routine clinical analysis. LC is sensitive towards interfering compounds and the precipitation of analytes in the column [2]. In the last decade, microscale chromatographic separation techniques, such as L-LC-MS (20), have been designed to resolve these limitations—they permit easy column preparation, high permeability, low backpressure, fast analyte mass transfer, and versatile surface chemistry [21].

CE-MS uses free-flow separation of analytes in buffer-filled capillaries [22]. Therefore, it is relatively robust and compatible with most volatile buffers without continuous adjustment of the ionization voltage for optimal ESI, mainly due to the absence of buffer gradients, as is the case in LC-coupled MS analysis. Therefore, CE-MS certainly appears to be an excellent choice for fast high-resolution analysis of complex biological samples.

In contrast to the platforms already discussed above, detection of specific proteins with protein microarrays is a non-MS-based targeted proteomics approach. The technique imprints specific antibodies or antigens on a plate to facilitate immune detection of multiple proteins. A single sample is hybridized to the array followed by the detection of the captured antigens or antibodies [23]. However, the requirement for antibodies with accurate sequence information and with high specificity for target peptides limits usage in clinical analysis.

6.4 MS Techniques

A detailed overview of the different ionization processes and the recent advances in mass spectrometry is far beyond the scope of this review and will be introduced in another chapter.

Briefly, ionization of polypeptides can technically be performed by electrospray ionization (ESI) or matrix-assisted laser desorption ionization (MALDI). ESI generates charged droplets in a high-voltage field. In this field, the solvent in the droplet evaporates, generating multiply charged analyte ions, which make it possible to analyze high-molecular-weight proteins. ESI pre-MS separation can be coupled online with a physical connection to an MS. This approach may be less stable compared with MALDI due to possible electrospray collapse, but it is less susceptible to the signal suppression phenomenon, in which certain analytes are preferentially ionized, and hence easily detected, but other analytes may become masked from detection [24].

For MALDI, the sample is coated onto a target plate with a matrix and co-crystallized. By absorbing the energy of the laser, the matrix desorbs the analyte. For proteomic applications, samples separated after an entire LC or CE run is spotted off-line, i.e., without mechanical connection to the MS. Off-line coupling is technically less demanding, but tends to result in signal loss.

Using MS approaches, peptide sequences of candidate biomarkers can be determined with two MS devices connected to each other (tandem MS or MS/MS). In the first step, MS isolates the ions of interest (the parent ions). In the second step, the parent ions are subjected to fragmentation by two different dissociation steps, by collision with other molecules (collision-induced dissociation) or by the transfer of electrons to the target molecule (electron transfer dissociation). The resulting fragments (the daughter ions) are analyzed using the second MS instrument. The MS/MS spectrum of the fragments displays the sequence of the parent ion, which is routinely searched against databases to identify the protein biomarker from which it originated.

As an alternative to the label-free approaches, chemical labeling of proteins/peptides prior to fractionation and MS analysis is frequently used. There are a large number of available labeling strategies. Isobaric tagging for relative and absolute quantitation (iTRAQ), tandem mass tagging (TMT) [25], isotope coded affinity tagging (ICAT) [26], and isotope coded protein labeling (ICPL) [27], are among the most popular commercial alternatives, while dimethylation and $^{18}O/^{16}O$ are among the non-commercial alternatives. All these methods have advantages and disadvantages. iTRAQ (4-plex and 8-plex) and TMT (6-plex) label free amines that are found in all peptides. Reporter ions with different masses that relate to the different samples are released from the peptide during MS/MS fragmentation and are used for quantification.

6.5 Data Analysis

The analysis of proteomic data requires adequate tools to handle its large scale and complexity. The essentials to be considered are linearity and reproducibility of the quantity of polypeptides. For in-gel analysis, fluorescent dyes provide larger linear dynamic ranges than conventional protein stains such as Coomassie Blue or silver stains and are now routinely used for quantification, with similar or better sensitivity.

For MS-based proteomics, relative quantification based on ion counting is straightforward, but absolute quantification is usually achieved using an isotope-labeled marker analog. For comparative data evaluation, it is essential to identify identical polypeptides with high probability in consecutive samples. Hence, the resolution and accuracy of the parameters used for target definition are of major importance, such as retention time for LC separation, migration time for CE separation, and molecular mass. Separation-based parameters often vary with the ion content of the samples, requiring thorough data normalization. Adequate approaches have been developed using either external standards or polypeptides that are found with high frequency in the analyzed samples and that serve as internal standards [23]. The normalization has to compensate for deviations in all three dimensions of data; the two defining parameters, e.g., retention time and molecular weight, and signal intensity.

Statistical data mining when dealing with biological samples is often influenced by the fact that the variables highly outnumber the available samples. This generally results in a higher dimension of mathematical problem solving, and the selection of significant items has to be considered through respective adjustments for multiple hypothesis testing.

In bioinformatics for the proteomic analysis of neurodegenerative diseases, medical definitions of disease are often made using generic descriptions of observed symptoms rather than the exact pathological processes involved. Thus, there is significant overlapping, often making it difficult to distinguish between AD, PD, and aging. Consequently, the concept of a single biomarker, allowing not only diagnosis but also staging and prognosis, appears questionable. Combining multiple biomarkers into a diagnostic or predictive pattern may help address this problem.

6.6 Aging and Dementia: General Aspects

In the preceding section, we attempted to provide a concise overview of the technical aspects of proteomics study. In the following section, we will present some examples demonstrating the general principles of clinical proteomics and discuss potential future developments in the study of neurodegenerative diseases and age-related process. In addition, we will mention our novel proteomics approach for examining body fluids. We will start with the process of aging itself and proceed to a discussion of neurodegenerative diseases, focusing on cognitive disorders such as AD.

The aging process can be understood as a multidimensional change in an organism over time, in physical, psychological, and social aspects. In humans, telomere shortening occurs in the vast majority of tissues during normal aging and is accelerated in chronic diseases with increased cellular turnover. Abnormal telomere function interferes with the proliferative capacity of human cells and promotes senescence and/or apoptosis. Recently, researchers identified protein markers secreted from telomere-dysfunctional bone marrow cells [28]. The expression levels of these proteins—CRAMP, stathmin, EF-l, and chitinase—were increased in blood and in various other tissues in aging telomere-knockout mice, but not in aging mice with long telomeres. These proteins are upregulated in late passage pre-senescent human fibroblasts. Increases in these marker proteins were also observed in blood plasma of aging humans. Moreover, there was a significant increase in the expression of these biomarkers in the blood plasma of patients with chronic diseases, probably because of increased rates of cell turnover and telomere shortening. These results support the concept that telomere dysfunction is connected to the pathways that are activated during human aging and disease, as well as the associated DNA damage. Furthermore, both telomere dysfunction and DNA damage may decrease the fidelity of protein synthesis [29], promoting age-related damage and changes in cellular morphology. Therefore, it is important to understand and detect changes in physiological function and structure that normally occur with aging, and to distinguish these changes from alterations that take place in disease.

6.7 The Aging Brain

Cognitive performance is attenuated with age in both humans and animal models, even in the absence of overt pathological changes. For many years, diminished cognitive function was believed to be a reflection of neuronal loss throughout the lifespan, and neuronal loss beyond a certain threshold was thought to produce symptoms of dementia. Recent studies, however, utilizing contemporary stereological quantification techniques, have revealed that such neuronal loss rarely occurs during the course of normal, healthy aging [30]. Therefore, the age-related decline in cognitive function must predominantly depend on dysregulated and inefficient neural processing.

Although the underlying mechanisms for the decrease in neural processing as the brain ages remain unknown, age-related changes in glucose metabolism, protein expression (synaptophysin, calcium regulatory proteins, excitatory and inhibitory neurotransmitter receptors, etc.), neurotransmitter levels (dopamine, norepinephrine, 5-HT, GABA, acetylcholine, etc.), as well as in trophic factor support (brain-derived neurotrophic factor, nerve growth factor, insulin-like growth factor-1, etc.) [31–40] have been demonstrated. At the cellular level, electrophysiological correlates of learning and memory have been demonstrated. Considerable evidence also exists for age related changes in dendritic and synaptic morphology, as well as in vascular density [5, 41].

The application of proteomic analyses to neurobiological studies of aging is becoming more common [42, 43] as the field adopts more systems-oriented approaches. For example, proteomic studies of the hippocampus in aging have revealed regional dysregulation of various processes, including metabolism, glutamate processing and protein synthesis [44, 45], cytoprotection and apoptosis [46], and synaptic integration and function [47].

Phosphatidylethanolamine binding protein (PEBP or hippocampal cholinergic neurostimulating peptide [HCNP]) has been previously reported to decrease with age in the human hippocampus [48]. Dysregulated expression of coronin 1A, an F-actin binding protein involved in the regulation of the cytoskeleton [49], is associated with alterations in hippocampal structure and reductions in the number of synapses in this brain region [50, 51]. Levels of protein phosphatase 1 (PP1), a modulator of learning and memory [52], also significantly decrease with age, while the expression of the vesicular fusion protein NSF increases. Expression of peroxiredoxin 6, a potent antioxidant and free radical scavenger found in hippocampal neuroglia [53], was increased in old rats, possibly in response to increased oxidative stress [54–56]. Interestingly, increased peroxiredoxin expression has been reported in brain in neurodegenerative disorders characterized by elevated oxidative stress. In contrast, levels of glutathione synthetase, which catalyzes the second step in the biosynthesis of the antioxidant glutathione, diminish with aging, providing further evidence for increased oxidative damage in the aged brain [57, 58].

6.8 Neurobiology of Aging and Cognitive Decline

Although the etiology of age-related cognitive decline is unclear, it is associated with impaired functioning of many cellular processes, e.g., reduced blood flow and glucose metabolism, as well as alterations in synaptic morphology and electrophysiological correlates of learning and memory [57, 58]. Previous studies of gene expression during aging identified cognitive status-associated gene expression patterns related to synaptic morphology, transcription factor expression, protein modification [59], glucose utilization, and signal transduction processes [39]. Proteomic reports have addressed age-associated changes in hippocampal protein expression and posttranslational modification [44, 60, 61]. Other reports focused on specific proteins related to synaptic transmission [51, 62, 63], mitochondrial function, oxidative stress [44, 55, 64], as well as proteolytic proteins [45, 65].

Specific changes associated with cognitive status have been investigated, and significant increases in two heat-shock 70 family members—heat-shock protein 4 and heat-shock protein 105/110 kDa—emerged only when a cognitive status-specific comparison was made. The function of these chaperones is to ensure proper folding of newly synthesized proteins and to prevent protein aggregation. Heat-shock proteins are induced by protein misfolding and oxidative stress, and have been demonstrated to increase in both aged rat brain [46, 66, 67] and human hippocampus in AD [68].

Alpha enolase was significantly decreased in old rats compared with young rats, but was significantly increased in old impaired rats compared with old non-impaired rats. These data suggest that alpha enolase, an essential glycolytic enzyme [69], is differentially regulated by normal aging and age-related cognitive decline. Aldolase, phosphofructokinase, phosphoglycerate mutase, and malate dehydrogenase, additional metabolic enzymes, were also elevated in cognitively impaired senescent animals compared with non-impaired animals of any age. This is consistent with observations that glycolytic enzymes increase in the aged mouse brain, as a compensatory response to decreased metabolic activity [70]. Reports of altered metabolism in age-related cognitive decline [39, 71, 72] are provided below.

Oxidative stress is a well-established characteristic of the aged brain and has been implicated in cognitive impairment and hippocampal dysfunction [60, 73]. Upregulation of proteins involved in oxidative carbohydrate metabolism in the hippocampus of old, cognitively impaired rats may contribute to the generation of reactive oxygen species and free radicals, thereby increasing oxidative stress. However, oxidative inactivation of proteins involved in carbohydrate metabolism may decrease energy metabolism and prevent neuronal damage [74]. Human studies demonstrate that cognitive decline is closely related to impaired cerebral blood flow and metabolism, while only modest neuronal loss occurs [75, 76]. Increased oxidative stress and metabolic derangements likely contribute to neuronal dysfunction [77, 78]. The combined insults of decreased cerebral blood flow, chronic hypoxia, and decreased glucose utilization in aged rats could upregulate glycolytic enzymes in the hippocampus as a compensatory response that further contributes to oxidative stress and increased levels of stress proteins, thereby exacerbating cognitive decline in a subset of animals.

Future studies of proteomic changes with cognitive decline and/or age will require additional technical refinements. While the changes described here are of interest, they represent fairly abundant protein species. Specific strategies to increase sensitivity to permit detection of low abundance proteins will have to be applied in future studies, such as synaptosomal preparations [79, 80] and affinity trapping for specific post-translational modifications. Several excellent reports by Butterfield and colleagues [44, 64, 70, 74] demonstrate the importance of examining posttranslational modifications and their effects on protein function in the aged hippocampus.

6.9 Proteomics of CSF and Other Body Fluids

CSF has been extensively studied because it provides an important source of potential biomarkers for brain disorders and therapeutic drug development strategies [81]. Studies on the CSF could provide potential biomarker information for neurological studies and increase our general knowledge of a number of human diseases.

However, there are still challenges in using CSF and plasma as sample sources for neuroproteomic studies [82, 83]. CSF can be obtained through lumbar puncture, which is frequently performed in the differential diagnosis of neurological disorders. In addition, most neurological diseases are related to processes that occur in the body as a whole, such as inflammation. Therefore, to define disease specific CSF/plasma proteins, one must first demonstrate the specificity of proteins linked to inflammation or other non-neuronal-specific reactions [84]. Furthermore, the difference among individuals or even in the same person at different CSF and plasma sample collection time points will also decrease the usefulness of the CSF/plasma protein multivariate marker. Accordingly, post-neuroproteomic validation using different methods is extremely important for confirming the specificity of disease-related proteins.

The second challenge towards establishing better CSF/plasma protein markers is to identify low-abundance proteins with important functions, especially those proteins with rapidly changing expression, as well as proteins for which expression is heavily temporally and spatially regulated. To meet this challenge, quantitative proteomic approaches should be combined with a number of real-time analysis techniques, which requires more effort and a large amount of work for peptide identification.

Protein diversity is also a challenge when using CSF and plasma as input sources for neuroproteomics. Protein diversity can include factors such as post-translational modification, as well as splicing variants. Although proteomic approaches for identifying splice variants at the protein level are still under development, current research has been encouraging, as shown in a recent report by Blakeley [85].

6.10 Neuroproteomics in the Study of CSF in Neurodegenerative Disorders

There has been rapid progress in the application of proteomics to the study of neurodegenerative disorders [86]. CSF samples from patients with AD [82], Down's syndrome, Pick's disease, PD [87, 88], and other disorders have been analyzed by proteomics to identify biomarkers to aid in diagnosis or to monitor disease progression [89, 90]. As expected, most of these studies aimed to find biomarkers for early diagnosis, as well as to discover key proteins that are functionally important during disease progression. In addition, the differentially expressed proteins identified by proteomic approaches are also important for examining the effects of drug treatment. Therefore, as expected, neuroproteomics has been an important approach in uncovering the molecular basis of various neurological diseases in order to understand their pathogenesis [91, 92]. This field has been termed clinical proteomics [93] and is used not only in AD but also in major depression [83], schizophrenia [83, 94, 95], and multiple sclerosis [96, 97]. For example, CSF biomarkers such as total tau, phosphorylated tau, and the 42-amino acid form of β-amyloid are commonly studied examples, as described by Blennow *et al.* [82]. Serum biomarkers for neurological diseases, such as AD, have also attracted a lot of attention in proteomic studies, and it is important to distinguish AD from controls [98, 99]. Protein changes in AD, such as the accumulation of misfolded neuronal proteins, amyloid and tau, into insoluble aggregates known as extracellular senile plaques and intracellular neurofibrillary tangles, have provided a good model for the use of the proteomic approach [89].

In addition to targeted proteomics, which has been used in the study of AD by focusing on β-amyloid itself and its presence in biological fluids, efforts have also focused on identifying clinically useful markers of disrupted β-amyloid homeostasis in AD [84] and improving its efficiency for clinical studies. In addition, the use of quantitative proteomics has also contributed to the identification of AD-related proteins during recent years. Compared with age-matched controls, quantitative proteomics from the CSF from AD patients, as well as from patients with other neurodegenerative diseases, will allow us to identify disease related proteins that may serve as specific biomarkers for AD and other geriatric dementias [100].

Several proteomic studies on AD have been conducted, and many of them are summarized in review papers on AD and other neurodegenerative diseases [93, 101]. In a recent update, Zetterberg *et al.* [102] listed 30 CSF proteins that were identified in two or more proteomic studies as possible biomarkers for AD. Similarly, Korolainen *et al.* [93] presented a list of 26 proteins that displayed statistically significant changes in AD.

Some proteomics studies have used SELDI-TOF-MS [103, 104] and 2D-PAGE [105] for the detection of differentially expressed proteins between AD patients and controls. 2D-DIGE has also been used to detect intra and inter-individual differences among patients with mild dementia, but no cognitive disease [106]. Abdi *et al.* [107] identified more than 1500 proteins using the iTRAQ labeling approach and found that 136 proteins were uniquely changed in CSF from AD compared with CSF from PD, dementia with Lewy bodies and controls. Yin *et al.* combined 1D-PAGE and 2D-PAGE with LC-MS/MS and detected 21 proteins that had different abundance levels between AD and controls [108].

In addition to routine proteomics, redox neuroproteomics was also used to study the AD brain in order to uncover the effect of oxidative stress in AD patients. Oxidative stress was specifically studied because of the critical role of β-amyloid peptide and oxidative stress in AD pathogenesis and progression [109] and in PD [60]. The information obtained from redox proteomics may not only be helpful in understanding the molecular mechanisms involved in the development and progression of AD and other neurodegenerative disorders, but it may also provide potential targets for drug therapy [60]. By using the redox proteomics approach, Sultana R *et al.* identified a number of commonly oxidized proteins in different brain regions in the AD brain and suggested a potential role for these oxidized proteins, and thus, oxidative stress, in the pathogenesis of AD [110]. Similarly, Butterfield *et al.* reported that protein modification by oxidation is an important common feature in both familial and sporadic AD [74]. The application of redox proteomics to study cellular events, especially those related to disease dysfunction, may provide an efficient tool for understanding the main mechanisms involved in the pathogenesis and progression of oxidative stress-related neurodegenerative disorders [111].

PD is currently diagnosed based on clinical symptoms [112] that include resting tremor, muscle rigidity, and bradykinesia. Clinical symptoms develop after loss of 70–80% of dopaminergic neurons in the substantia nigra [113, 114]. About 2% of the population over 65 years is affected by PD. More than 90% of PD cases are sporadic and associated with unknown causes, while 10% represent familial inherited forms resulting from mutations in genes. PD is characterized by the occurrence of Lewy bodies, mainly consisting of heavily ubiquitinated α-synuclein [115], which is normally present in the presynaptic terminals of most neurons. Impairment in lysosomal protein degradation and protein aggregation influences the fibrillization of α-synuclein [116]. The toxicity of the insoluble oligomers of α-synuclein is unclear, whereas the intermediates generated during fibrillization are thought to be toxic [117]. α-Synuclein is detectable in blood and CSF as both forms. α-synuclein among other cofactors induces fibrillization of tau [118, 119]. Neurofibrillary tangles of tau are one of the hallmarks of AD. The tau pathology may be part of a final common pathway for neurodegeneration where an increased level of total tau in CSF reflects neuronal damage. A few proteomic studies have been performed to discover new potential biomarkers for PD, with the most common method being 2D-PAGE [98, 108, 120–122]. Two different quantitative proteomic approaches were used to compare CSF from AD, PD and control groups. In a study using iTRAQ labeling, Abdi *et al.* identified 73 proteins with differential expression in PD patients [107], whereas Pan *et al.* used SCX in combination with LC-MS/MS and the addition of stable isotope-labeled peptides to absolutely quantify five proteins.

If neuroproteomics can be appropriately tailored to examine protein changes in complex neurological diseases resulting from multiple factors, both genetic and environmental, in a variety of tissues and biological fluids, from post-mortem CNS tissue samples, to cerebrospinal fluid and blood, as well as in mutant animal models with neurodegenerative disease phenotypes, it has the potential to identify numerous markers and more efficiently advance clinical and basic research. We anticipate that proteomics coupled with genomics and bioinformatics will aid in discovering biomarkers for neurodegeneration.

6.11 New Approach in Clinical Neuroproteomics: Protein Co-Expression Analysis of Human Samples from Brain, CSF, Serum, and Urine

6.11.1 Introduction

Various body fluids, including blood (serum or plasma) and CSF, have been used in neuroproteomics to seek out disease-specific and surrogate markers. An emerging technique for comparative proteomics is 2D-DIGE; it improves the reproducibility and reliability of differential protein expression analysis between samples [84, 86]. In this method, different samples are labeled with spectrally resolvable fluorescent dyes—Cy2, Cy3 or Cy5—and mixed prior to IEF and resolved on the same 2DE gel. The fluorescent dyes have very similar molecular masses and are positively charged to match the charge that is replaced on the lysine residue. The charge and mass matching ensure that all the samples essentially co-migrate during electrophoresis. This method is generally used for comparison between test (i.e., disease) and control samples. However, in our novel strategy, we have used the technique to identify proteins that are co-expressed in brain and body fluids. We conjectured that a set of co-expressed proteins may provide a biomarker panel for neurodegenerative disease processes. In this study, we discovered a number of co-expressed proteins that are present in both brain and body fluids such as serum, CSF and urine.

6.11.2 Materials and Methods

6.11.2.1 Samples

In this study, we used human samples, including brain, serum, CSF, and urine from patients with no diagnosis of neurodegenerative disease. Furthermore, we ensured that the brain tissue samples had no clinical or morphologic evidence of pathology.

6.11.2.2 Sample preparation

Brain tissue was thoroughly sonicated in lysis buffer (30 mM Tris, 7 M urea, 2 M thiourea, 4% (w/v) CHAPS, pH 8.5) using a hand sonicator and centrifuged at 100,000g for 1 h at 15°C. The supernatant was collected and protein concentration was measured using the Bradford method.

Serum samples were collected and immediately centrifuged at 2500g for 5 min and supernatants were collected. Albumin and IgG were removed using an albumin and IgG removal kit (Amersham Biosciences Sunnyvale, CA) according to the manufacturer's protocol. Protein samples were concentrated by acetone precipitation. CSF samples, stored at −80°C, were slowly thawed on ice and immediately concentrated by acetone precipitation. Urine samples were collected and immediately centrifuged at 2000g for 5 min, and the resulting supernatants were filtered using filter paper on ice. Filtered samples were concentrated using a Centricon Plus-20 concentrator (Millipore, Bedford, MA, USA) and acetone precipitation. After acetone precipitation, samples were solubilized in lysis buffer (same as brain) using a hand sonicator and centrifuged at 100,000 g for 1 h at 15°C. The supernatants were collected and protein concentrations were measured using the Bradford method.

6.11.2.3 2DE gel electrophoresis

Prior to isoelectric focusing (IEF, the first dimension of 2DE gel electrophoresis), labeled samples, to be separated in the same gel, were mixed and added to DTT and Pharmalyte pH 3–10 (final concentration was 65 mM and 1% (v/v), respectively).

First-dimension electrophoresis was carried out using an immobilized pH gradient gel (immobilized dry strip gel, pH 4–7, 180 × 3 × 0.5 mm) with a horizontal electrophoresis apparatus (Multiphor II, Amersham Biosciences, Sunnyvale, CA). The sample was applied to the anodic end of the gel. IEF was performed using the following parameters: 0–200 V for 5 min, 200–300 V for 5 h, 300–3 500 V for 8 h, and 3 500 V for 8 h. Second-dimension electrophoresis was carried out on a 12–14% gradient polyacrylamide gel (ExcelGel XL SDS 12–14, 245 × 180 × 0.5 mm) according to the manufacturer's protocol.

6.11.2.4 Gel scanning and image analysis

Labeled proteins were visualized using the Typhoon™ 9400 imager (Amersham Biosciences, Sunnyvale, CA). Cy2 images were scanned using a 488 nm laser and an emission filter of 520 nm/BP (band pass) 40, and Cy3 images were scanned using a 532 nm laser and an emission filter of 580 nm. All gels were scanned at 100 nm resolution.

We examined overlapping spots on the same gel using ImageQuant V 5.2 (Amersham Biosciences, Sunnyvale, CA). Cy2-labeled proteins were displayed green and Cy3-labeled proteins were displayed red, and overlapping co-expressed proteins were displayed yellow.

6.11.2.5 Protein identification by mass spectrometry and database analysis

Mass spectrometric identification of proteins was performed as previously described. Briefly, proteins were excised from gels, followed by in-gel digestion with trypsin (Promega, Madison, WI, USA) in a buffer containing 50 mM ammonium bicarbonate (pH 8.0) and 2% acetonitrile overnight at 37°C. Molecular mass analysis of tryptic peptides was performed with matrix-assisted laser desorption/ionization time-of-flight mass spectrometry (MALDI-TOF/MS) using an ultraflex TOF/TOF (Bruker Daltonics, Billerica, MA, USA). Proteins were identified by comparison between the molecular weights determined by MALDI-TOF/MS and theoretical peptide masses of the proteins registered in NCBInr.

6.11.3 Results

6.11.3.1 Identification of proteins co-expressed in brain and serum

In order to identify proteins expressed in both human brain and serum, we combined 2DE and CyDye labeling—brain samples were labeled with Cy3 and serum samples were labeled with Cy2. We removed albumin and IgG from serum samples by using an albumin and IgG Removal kit (Amersham Biosciences) prior to labeling. After 2DE, we identified overlapping spots and selected 26 of them. We analyzed only 17 spots, because nine could not be silver stained, and therefore, could not be excised for MS. It may be possible that protein-staining techniques with higher sensitivity may help increase the number of co-expressed proteins available for MS analysis.

Contrary to the expectation, MS analysis indicated that only two proteins were co-expressed (apolipoprotein A-I and albumin). Results are shown in Figs. 6.1 and 6.2 and in Tables 6.1, 6.2 and 6.3.

6.11.3.2 Identification of proteins co-expressed in brain and CSF

After 2DE, we found that 16 pairs of spots overlapped. We analyzed only 14 of them because two could not be silver stained. MS analysis revealed that four proteins were co-expressed (apolipoprotein A-I, peroxiredoxin 2 isoform a, neuronal cell adhesion molecule and albumin). Results are shown in Figs. 6.1 and 6.2, and in Tables 6.2 and 6.3.

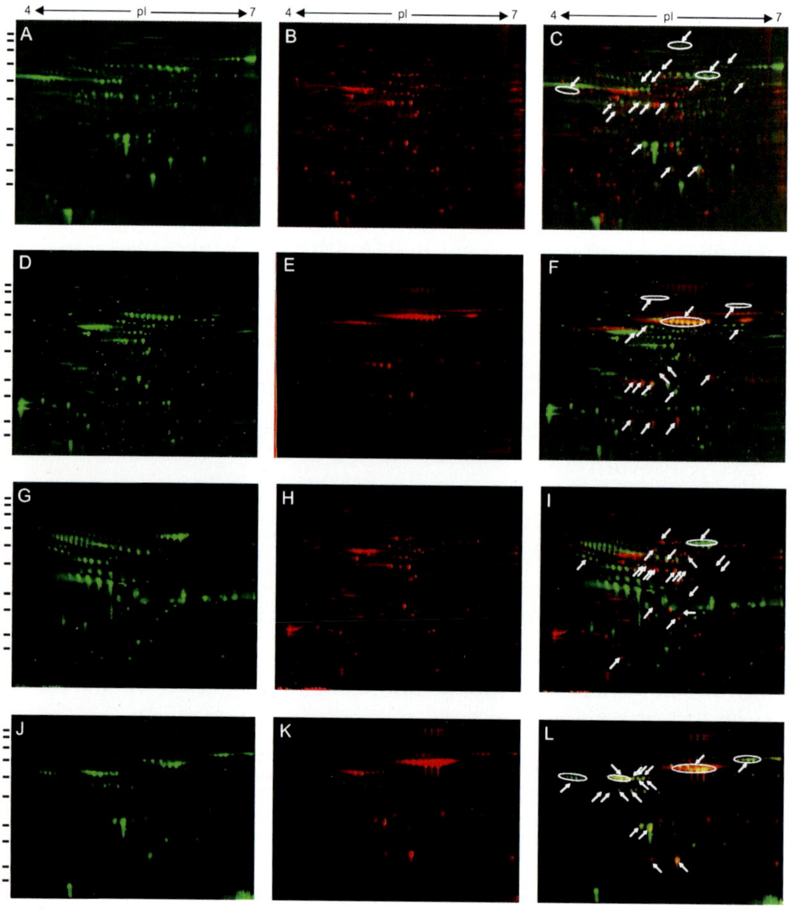

Figure 6.1 Fluorescence images of 2DE gels of protein co-expression patterns between the following samples: (A) serum, (B) brain and (C) merge; (D) brain, (E) CSF and (F) merge; (G) urine, (H) brain and (I) merge; (J) serum, (K) CSF, and (L) merge. All samples were 50 μg protein. All gels pH 4–7 IPG. Molecular weight markers indicate 250, 150, 100, 75, 50, 37, 25, 20, 15, and 10 kDa, starting from the top. Arrows indicate overlapping spots among the gels.

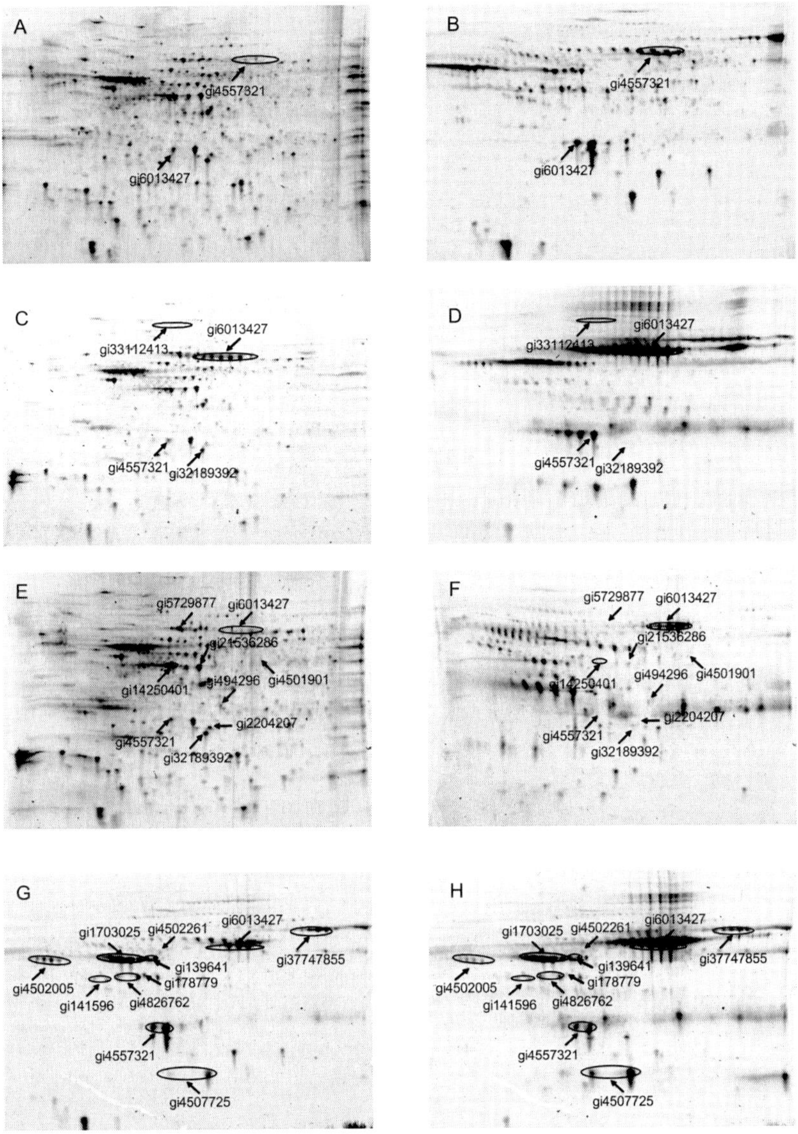

Figure 6.2 Image of 2DE gels of identified co-expressed proteins between the samples: (A) serum, (B) brain and (C) brain, (D) CSF and (E) urine, (F) brain and (G) serum, and (H) CSF. Arrows indicate spots containing co-expressed proteins among the gels, as determined by MS analysis.

Table 6.1 Cy dye labeling patterns between each samples

Cy dye	Brain & serum	Brain & CSF	Brain & urine	Serum & CSF
Cy2	Brain	Brain	Brain	Serum
Cy3	Serum	CSF	Urine	CSF

Table 6.2 Numbers of overlapping spots and MS analysis

Samples	No. of overlapping spots	No. of MS analysis
Brain & serum	26	17
Brain & CSF	16	14
Brain & urine	33	20
Serum & CSF	26	17

Note: The reason why it is different number between overlapping spots and MS analysis is that some spots could not be stained by silver staining for MS analysis and could not be excised. Therefore, it is possible that co-expression proteins increase by using higher sensitive silver staining for MS analysis.

Table 6.3 Identified co-expressed proteins between each samples by MS analysis

Accession number	Protein name	Relevance to AD
Brain & serum		
gi4557321	Apolipoprotein A-I	↓(CSF proteome)
gi6013427	Serum albumin	↓(CSF proteome)
Brain & CSF		
gi4557321	Apolipoprotein A-I	↓(CSF proteome)
gi32189392	Peroxiredoxine 2 isoform a	↑(brain proteome)
gi33112413	Neuronal cell adhesion molecule	
gi6013427	Serum albumin	↓(CSF proteome)
Brain & urine		
gi4557321	Apolipoprotein A-I	↓(CSF proteome)
gi32189392	Peroxiredoxine 2 isoform a	↑(brain proteome)
gi2204207	Glutathione S-transferase	
gi494296	Athepsin D	
gi14250401	Actin	
gi21536286	Brain creatine kinase	↑(brain proteome)
gi4501901	Aminoacylase 1	

Accession number	Protein name	Relevance to AD
gi5729877	Heat shock 70kDa protein 8	
gi6013427	Serum albumin	↓(CSF proteome)
Serum & CSF		
gi4507725	Transthyretin	↑ ↓ (CSF proteome)
gi4557321	Apolipoprotein A-I	↓(CSF proteome)
gi141596	Zinc-alpha-2-glycoprotein	
gi4826762	Haptoglobin	
gi1703025	Alpha-1-antitrypsin	↑(CSF proteome)
gi139641	Vitamin D-binding protein	
gi4502005	Alpha-2-HS-glycoprotein	↓(CSF proteome)
gi37747855	Transferrin	↓(CSF proteome)
gi6013427	Serum albumin	↓(CSF proteome)
gi4502261	Antithrombin III	
gi178779	Apolipoprotein A-IV	

Note: The list shows the NCBInr accession numbers and full names of the identified biomarkers, and their relevance to AD by reference citations.

6.11.3.3 Identification of proteins co-expressed in brain and urine

We sought to identify proteins expressed in both brain (labeled with Cy3) and urine (labeled with Cy2). We concentrated urine with a Centricon Plus-20 concentrator (Millipore) and acetone precipitation. After 2DE, we found that 33 pairs of spots overlapped. We analyzed only 20 because the others did not stain with silver. Surprisingly, MS analysis demonstrated that nine proteins were co-expressed (apolipoprotein A-I, peroxiredoxin 2 isoform a, glutathione S-transferase, cathepsin D, actin, brain creatine kinase, aminoacylase 1, heat-shock 70 kDa protein 8 and albumin). Results are shown in Figs. 6.1 and 6.2 and in Tables 6.2 and 6.3.

6.11.3.4 Identification of proteins co-expressed in serum and CSF

We anticipated identifying numerous proteins co-expressed in serum and CSF, because both 2DE images were very similar. Using

2DE, we found that 26 pairs of spots overlapped. We analyzed only 17 because the others could not be detected with silver staining. MS analysis revealed that 11 proteins were co-expressed (apolipoprotein A-I, transthyretin, zinc-alpha-2-glycoprotein, haptoglobin, alpha-1-antitrypsin, vitamin D-binding protein, alpha-2-HS-glycoprotein, transferrin, albumin, antithrombin III, and apolipoprotein A-IV). Results are shown in Figs. 6.1 and 6.2 and Tables 6.2 and 6.3.

6.12 Conclusions and Outlook

The 2D-DIGE method, using CyDye labeling, has been useful for identifying proteins that are differentially expressed between disease and control samples. We altered the strategy and used 2D-DIGE to detect proteins co-expressed in different human samples; for example, brain and serum. The aim was to identify proteins (including isoforms) co-expressed in both brain and body fluids to establish a panel of biomarkers for various neurodegenerative diseases.

A rapid increase in disease-related proteomic information is expected in the very near future, which should permit broad clinical application. In contrast to the high expectations generated by more than 25 years of technological progress, the impact of proteomics on our understanding of the pathogenesis, diagnosis and treatment of disease is still modest. Essentially, all clinical proteomic investigations indicate that the optimal approach is to combine several disease-associated markers into a comprehensive biomarker panel. This approach should allow for effective disease diagnosis and staging, and help the clinician to evaluate disease progression and determine prognosis. However, a thorough validation of these markers in well-designed blinded studies is essential, and should, therefore, be mandatory.

To date, proteomics is still far from realizing its full potential in clinical settings, particularly in neurology, and to take full advantage of this method, further standardization is needed to enable meta-analysis of datasets from different laboratories. With this and other improvements, proteomics promises to enhance our fundamental understanding of disease processes and advance treatment, ultimately helping improve the quality of life for patients.

References

1. J Choudhary, SG Grant (2004) Proteomics in postgenomic neuroscience: the end of the beginning. *Nat Neurosci* 7: 440–445.
2. R Bakry, CW Huck, M Najam-ul-Haq, M Rainer, GK Bonn (2007) Recent advances in capillary electrophoresis for biomarker discovery. *J Sep Sci* 30: 192–201.
3. EC Andrade, DD Krueger, AC Nairn (2007) Recent advances in neuroproteomics. *Curr Opin Mol Ther* 9: 270–281.
4. M Svensson, K Sköld, A Nilsson, M Fälth, K Nydahl, P Svenningsson, *et al.* (2007) Neuropeptidomics: MS applied to the discovery of novel peptides from the brain. *Anal Chem* 79: 14–21.
5. DL Dickstein, D Kabaso, AB Rocher, JI Luebke, SL Wearne, PR Hof (2007) Changes in the structural complexity of the aged brain. *Aging Cell* 6: 275–284.
6. MS Chong, S Sahadevan (2005) Preclinical Alzheimer's disease: diagnosis and prediction of progression. *Lancet Neurol* 4: 576–579.
7. CM Connell, MR Janevic, MP Gallant (2001) The costs of caring: impact of dementia on family caregivers. *J Geriatr Psychiatry Neurol* 14: 179–187.
8. KD van Dijk, CE Teunissen, B Drukarch, CR Jimenez, HJ Groenewegen, HW Berendse, *et al.* (2010) Diagnostic cerebrospinal fluid biomarkers for Parkinson's disease: a pathogenetically based approach. *Neurobiol Dis* 39: 229–241.
9. GS Omenn, DJ States, M Adamski, TW Blackwell, R Menon, H Hermjakob, *et al.* (2005) Overview of the HUPO plasma proteome project: results from the pilot phase with 35 collaborating laboratories and multiple analytical groups, generating a core dataset of 3020 proteins and a publicly-available database. *Proteomics* 5: 3226–3245.
10. S Schaub, J Wilkins, T Weiler, K Sangster, D Rush, P Nickerson (2004) Urine protein profiling with surface-enhanced laser-desorption/ionization time-of-flight mass spectrometry. *Kidney Int* 65: 323–332.
11. D Theodorescu, S Wittke, MM Ross, M Walden, M Conaway, I Just, *et al.* (2006) Discovery and validation of new protein biomarkers for urothelial cancer: a prospective analysis. *Lancet Oncol* 7: 230–240.
12. GM Fiedler, S Baumann, A Leichtle, A Oltmann, J Kase, J Thiery, *et al.* (2007) Standardized peptidome profiling of human urine by magnetic bead separation and matrix-assisted laser desorption/ionization time-of-flight mass spectrometry. *Clin Chem* 53: 421–428.

13. H Mischak, R Apweiler, RE Banks, M Conaway, J Coon, A Dominiczak, *et al.* (2007) Clinical proteomics: a need to define the field and to begin to set adequate standards. *Proteomics Clin Appl* 1: 148–156.

14. SD Patterson, R Aebersold (1995) Mass spectrometric approaches for the identification of gel-separated proteins. *Electrophoresis* 16: 1791–814.

15. TL Wu (2006) Two-dimensional difference gel electrophoresis. *Methods Mol Biol* 328: 71–95.

16. HJ Issaq (2001) The role of separation science in proteomics research. *Electrophoresis* 22: 3629–3638.

17. NP Munro, DA Cairns, P Clarke, M Rogers, AJ Stanley, JH Barrett, *et al.* (2006) Urinary biomarker profiling in transitional cell carcinoma. *Int J Cancer* 119: 2642–2650.

18. P Zerefos, J Prados, S Kossida, A Kalousis, A Vlahou (2007) Sample preparation and bioinformatics in MALDI profiling of urinary proteins. *J Chromatogr B Analyt Technol Biomed Life Sci* 853: 20–30.

19. T Kislinger, AO Gramolini, DH MacLennan, A Emili (2005) Multidimensional protein identification technology (MudPIT): technical overview of a profiling method optimized for the comprehensive proteomic investigation of normal and diseased heart tissue. *J Am Soc Mass Spectrom* 16: 1207–1220.

20. M Soldi, C Sarto, C Valsecchi, F Magni, V Proserpio, D Ticozzi, *et al.* (2005) Proteome profile of human urine with two-dimensional liquid phase fractionation. *Proteomics* 5: 2641–2647.

21. R Wu, L Hu, F Wang, M Ye, H Zou (2008) Recent development of monolithic stationary phases with emphasis on microscale chromatographic separation. *J Chromatogr A* 1184: 369–392.

22. P Schmitt-Kopplin, M Frommberger (2003) Capillary electrophoresis-mass spectrometry: 15 years of developments and applications. *Electrophoresis* 24: 3837–3867.

23. D Theodorescu, D Fliser, S Wittke, H Mischak, R Krebs, M Walden, *et al.* (2005) Pilot study of capillary electrophoresis coupled to mass spectrometry as a tool to define potential prostate cancer biomarkers in urine. *Electrophoresis* 26: 2797–2808.

24. TM Annesley (2003) Ion suppression in mass spectrometry. *Clin Chem* 49: 1041–1044.

25. A Thompson, J Schäfer, K Kuhn, S Kienle, J Schwarz, G Schmidt, *et al.* (2003) Tandem mass tags: a novel quantification strategy for

comparative analysis of complex protein mixtures by MS/MS. *Anal Chem* 75: 1895–1904.

26. SP Gygi, B Rist, SA Gerber, F Turecek, MH Gelb, R Aebersold (1999) Quantitative analysis of complex protein mixtures using isotope-coded affinity tags. *Nat Biotechnol* 17: 994–999.
27. A Schmidt, J Kellermann, F Lottspeich (2005) A novel strategy for quantitative proteomics using isotope-coded protein labels. *Proteomics* 5: 4–15.
28. H Jiang, E Schiffer, Z Song, J Wang, P Zürbig, K Thedieck, *et al.* (2008) Proteins induced by telomere dysfunction and DNA damage represent biomarkers of human aging and disease. *Proc Natl Acad Sci U S A* 105: 11299–11304.
29. LE Orgel (1973) Ageing of clones of mammalian cells. *Nature* 243: 441–445.
30. PR Hof, JH Morrison (2004) The aging brain: morphomolecular senescence of cortical circuits. *Trends Neurosci* 27: 607–613.
31. MG Baxter, KM Frick, DL Price, SJ Breckler, AL Markowska, LK Gorman (1999) Presynaptic markers of cholinergic function in the rat brain: relationship with age and cognitive status. *Neuroscience* 89: 771–779.
32. TD Smith, MM Adams, M Gallagher, JH Morrison, PR Rapp (2000) Circuit-specific alterations in hippocampal synaptophysin immunoreactivity predict spatial learning impairment in aged rats. *J Neurosci* 20: 6587–6593.
33. F Mora, G Segovia, A Del Arco (2008) Glutamate–dopamine–GABA interactions in the aging basal ganglia. *Brain Res Rev* 58: 340–353.
34. MP Gavilán, E Revilla, C Pintado, A Castaño, ML Vizuete, I Moreno-González, *et al.* (2007) Molecular and cellular characterization of the age-related neuroinflammatory processes occurring in normal rat hippocampus: potential relation with the loss of somatostatin GABAergic neurons. *J Neurochem* 103: 984–996.
35. M Majdi, A Ribeiro-da-Silva, AC Cuello (2007) Cognitive impairment and transmitter-specific pre- and postsynaptic changes in the rat cerebral cortex during ageing. *Eur J Neurosci* 26: 3583–3596.
36. TC Foster (2007) Calcium homeostasis and modulation of synaptic plasticity in the aged brain. *Aging Cell* 6: 319–325.
37. L Shi, MM Adams, MC Linville, IG Newton, ME Forbes, AB Long, *et al.* (2007) Caloric restriction eliminates the aging-related decline in NMDA and AMPA receptor subunits in the rat hippocampus and induces homeostasis. *Exp Neurol* 206: 70–79.

38. IK Hwang, KY Yoo, BK Jung, JH Cho, DH Kim, TC Kang, *et al.* (2006) Correlations between neuronal loss, decrease of memory, and decrease expression of brain-derived neurotrophic factor in the gerbil hippocampus during normal aging. *Exp Neurol* 201: 75–83.

39. WB Rowe, EM Blalock, KC Chen, I Kadish, D Wang, JE Barrett, *et al.* (2007) Hippocampal expression analyses reveal selective association of immediate-early, neuroenergetic, and myelinogenic pathways with cognitive impairment in aged rats. *J Neurosci* 27: 3098–3110.

40. JC Gant, MM Sama, PW Landfield, O Thibault (2006) Early and simultaneous emergence of multiple hippocampal biomarkers of aging is mediated by Ca_{2+}-induced Ca_{2+} release. *J Neurosci* 26: 3482–3490.

41. WE Sonntag, CD Lynch, PT Cooney, PM Hutchins (1997) Decreases in cerebral microvasculature with age are associated with the decline in growth hormone and insulin-like growth factor 1. *Endocrinology* 138: 3515–3520.

42. O Carrette, PR Burkhard, DF Hochstrasser, JC Sanchez (2006) Age-related proteome analysis of the mouse brain: a 2-DE study. *Proteomics* 6: 4940–4949.

43. HF Poon, A Castegna, SA Farr, V Thongboonkerd, BC Lynn, WA Banks, *et al.* (2004) Quantitative proteomics analysis of specific protein expression and oxidative modification in aged senescence-accelerated-prone 8 mice brain. *Neuroscience* 126: 915–926.

44. HF Poon, HM Shepherd, TT Reed, V Calabrese, AM Stella, G Pennisi, *et al.* (2006) Proteomics analysis provides insight into caloric restriction mediated oxidation and expression of brain proteins associated with age-related impaired cellular processes: mitochondrial dysfunction, glutamate dysregulation and impaired protein synthesis. *Neurobiol Aging* 27: 1020–1034.

45. M Paz Gavilán, J Vela, A Castaño, B Ramos, JC del Río, J Vitorica, *et al.* (2006) Cellular environment facilitates protein accumulation in aged rat hippocampus. *Neurobiol Aging* 27: 973–982.

46. V Calabrese, G Scapagnini, A Ravagna, C Colombrita, F Spadaro, DA Butterfield, *et al.* (2004) Increased expression of heat shock proteins in rat brain during aging: relationship with mitochondrial function and glutathione redox state. *Mech Ageing Dev* 125: 325–335.

47. E Head, MM Corrada, K Kahle-Wrobleski, RC Kim, F Sarsoza, M Goodus, *et al.* (2009) Synaptic proteins, neuropathology and cognitive status in the oldest-old. *Neurobiol Aging* 30: 1125–1134.

48. H Yuasa, K Ojika, S Mitake, E Katada, N Matsukawa, Y Otsuka, *et al.* (2001) Age-dependent changes in HCNP-related antigen expression in the human hippocampus. *Brain Res Dev Brain Res* 127: 1–7.

49. VE Galkin, A Orlova, W Brieher, HY Kueh, TJ Mitchison, EH Egelman (2008) Coronin-1A stabilizes F-actin by bridging adjacent actin protomers and stapling opposite strands of the actin filament. *J Mol Biol* 376: 607–613.

50. BP Rutten, NM Van der Kolk, S Schafer, MA van Zandvoort, TA Bayer, HW Steinbusch, *et al.* (2005) Age-related loss of synaptophysin immunoreactive presynaptic boutons within the hippocampus of APP751SL, PS1M146L, and APP751SL/PS1M146L transgenic mice. *Am J Pathol* 167: 161–173.

51. L Shi, MC Linville, EW Tucker, WE Sonntag, JK Brunso-Bechtold (2005) Differential effects of aging and insulin-like growth factor-1 on synapses in CA1 of rat hippocampus. *Cereb Cortex* 15: 571–577.

52. D Genoux, U Haditsch, M Knobloch, A Michalon, D Storm, IM Mansuy (2002) Protein phosphatase 1 is a molecular constraint on learning and memory. *Nature* 418: 970–975.

53. MH Jin, YH Lee, JM Kim, HN Sun, EY Moon, MH Shong, *et al.* (2005) Characterization of neural cell types expressing peroxiredoxins in mouse brain. *Neurosci Lett* 381: 252–257.

54. AN Donahue, M Aschner, LH Lash, T Syversen, WE Sonntag (2006) Growth hormone administration to aged animals reduces disulfide glutathione levels in hippocampus. *Mech Ageing Dev* 127: 57–63.

55. TC Squier (2001) Oxidative stress and protein aggregation during biological aging. *Exp Gerontol* 36: 1539–1550.

56. Y Zhu, PM Carvey, Z Ling (2006) Age-related changes in glutathione and glutathione-related enzymes in rat brain. *Brain Res* 1090: 35–44.

57. ES Rosenzweig, CA Barnes (2003) Impact of aging on hippocampal function: plasticity, network dynamics, and cognition. *Prog Neurobiol* 69: 143–179.

58. LD Brewer, O Thibault, J Staton, V Thibault, JT Rogers, G Garcia-Ramos, *et al.* (2007) Increased vulnerability of hippocampal neurons with age in culture: temporal association with increases in NMDA receptor current, NR2A subunit expression and recruitment of L-type calcium channels. *Brain Res* 1151: 20–31.

59. C Burger, MC López, JA Feller, HV Baker, N Muzyczka, RJ Mandel (2007) Changes in transcription within the CA1 field of the hippocampus are

associated with age-related spatial learning impairments. *Neurobiol Learn Mem* 87: 21–41.

60. DA Butterfield, M Perluigi, R Sultana (2006) Oxidative stress in Alzheimer's disease brain: new insights from redox proteomics. *Eur J Pharmacol* 545: 39–50.
61. O Weinreb, N Drigues, Y Sagi, AZ Reznick, T Amit, MB Youdim (2007) The application of proteomics and genomics to the study of age-related neurodegeneration and neuroprotection. *Antioxid Redox Signal* 9: 169–179.
62. MM Adams, L Shi, MC Linville, ME Forbes, AB Long, C Bennett, *et al.* (2008) Caloric restriction and age affect synaptic proteins in hippocampal CA3 and spatial learning ability. *Exp Neurol* 211: 141–149.
63. Y Sato, H Yamanaka, T Toda, Y Shinohara, T Endo (2005) Comparison of hippocampal synaptosome proteins in young-adult and aged rats. *Neurosci Lett* 382: 22–26.
64. HF Poon, V Calabrese, M Calvani, DA Butterfield (2006) Proteomics analyses of specific protein oxidation and protein expression in aged rat brain and its modulation by L-acetylcarnitine: insights into the mechanisms of action of this proposed therapeutic agent for CNS disorders associated with oxidative stress. *Antioxid Redox Signal* 8: 381–394.
65. M Nyffeler, WN Zhang, J Feldon, I Knuesel (2007) Differential expression of PSD proteins in age-related spatial learning impairments. *Neurobiol Aging* 28: 143–155.
66. K Unno, H Asakura, Y Shibuya, M Kaiho, S Okada, N Oku (2000) Increase in basal level of Hsp70, consisting chiefly of constitutively expressed Hsp70 (Hsc70) in aged rat brain. *J Gerontol A Biol Sci Med Sci* 55: B329–B335.
67. V Calabrese, G Scapagnini, C Colombrita, A Ravagna, G Pennisi, AM Giuffrida Stella, *et al.* (2003) Redox regulation of heat shock protein expression in aging and neurodegenerative disorders associated with oxidative stress: a nutritional approach. *Amino Acids* 25: 437–444.
68. R Sultana, D Boyd-Kimball, J Cai, WM Pierce, JB Klein, M Merchant, *et al.* (2007) Proteomics analysis of the Alzheimer's disease hippocampal proteome. *J Alzheimers Dis* 11: 153–164.
69. A Keller, A Bérod, M Dussaillant, N Lamandé, F Gros, M Lucas (1994) Coexpression of alpha and gamma enolase genes in neurons of adult rat brain. *J Neurosci Res* 38: 493–504.

70. HF Poon, RA Vaishnav, TV Getchell, ML Getchell, DA Butterfield (2006) Quantitative proteomics analysis of differential protein expression and oxidative modification of specific proteins in the brains of old mice. *Neurobiol Aging* 27: 1010–1019.

71. FH Gage, PA Kelly, A Björklund (1984) Regional changes in brain glucose metabolism reflect cognitive impairments in aged rats. *J Neurosci* 4: 2856–2865.

72. WE Sonntag, M Ramsey, CS Carter (2005) Growth hormone and insulin-like growth factor-1 (IGF-1) and their influence on cognitive aging. *Ageing Res Rev* 4: 195–212.

73. DB Miller, JP O'Callaghan (2003) Effects of aging and stress on hippocampal structure and function. *Metabolism* 52: 17–21.

74. DA Butterfield, A Gnjec, HF Poon, A Castegna, WM Pierce, JB Klein, *et al.* (2006) Redox proteomics identification of oxidatively modified brain proteins in inherited Alzheimer's disease: an initial assessment. *J Alzheimers Dis* 10: 391–397.

75. M Gallagher, JL Bizon, EC Hoyt, KA Helm, PK Lund (2003) Effects of aging on the hippocampal formation in a naturally occurring animal model of mild cognitive impairment. *Exp Gerontol* 38: 71–77.

76. PR Rapp, M Gallagher (1996) Preserved neuron number in the hippocampus of aged rats with spatial learning deficits. *Proc Natl Acad Sci U S A* 93: 9926–9930.

77. MM Esiri (2007) Ageing and the brain. *J Pathol* 211: 181–187.

78. S Hoyer (1982) The abnormally aged brain. Its blood flow and oxidative metabolism. A review — part II. *Arch Gerontol Geriatr* 1: 195–207.

79. CR Jiménez, M Eyman, ZS Lavina, A Gioio, KW Li, RC van der Schors, *et al.* (2002) Protein synthesis in synaptosomes: a proteomics analysis. *J Neurochem* 81: 735–744.

80. M Krüger, I Kratchmarova, B Blagoev, YH Tseng, CR Kahn, M Mann (2008) Dissection of the insulin signaling pathway via quantitative phosphoproteomics. *Proc Natl Acad Sci U S A* 105: 2451–2456.

81. MJ Romeo, V Espina, M Lowenthal, BH Espina, EF Petricoin, LA Liotta (2005) CSF proteome: a protein repository for potential biomarker identification. *Expert Rev Proteomics* 2: 57–70.

82. K Blennow, H Hampel, M Weiner, H Zetterberg (2010) Cerebrospinal fluid and plasma biomarkers in Alzheimer disease. *Nat Rev Neurol* 6: 131–144.

83. TJ Raedler, K Wiedemann (2006) CSF-studies in neuropsychiatric disorders. *Neuro Endocrinol Lett* 27: 297–305.

84. M Thambisetty, S Lovestone (2010) Blood-based biomarkers of Alzheimer's disease: challenging but feasible. *Biomark Med* 4: 65–79.

85. P Blakeley, JA Siepen, C Lawless, SJ Hubbard (2010) Investigating protein isoforms via proteomics: a feasibility study. *Proteomics* 10: 1127–1140.

86. A Westman-Brinkmalm, U Ruetschi, E Portelius, U Andreasson, G Brinkmalm, G Karlsson, *et al.* (2009) Proteomics/peptidomics tools to find CSF biomarkers for neurodegenerative diseases. *Front Biosci* 14: 1793–1806.

87. G Srivastava, K Singh, MN Tiwari, MP Singh (2010) Proteomics in Parkinson's disease: current trends, translational snags and future possibilities. *Expert Rev Proteomics* 7: 127–139.

88. KD van Dijk, CE Teunissen, B Drukarch, CR Jimenez, HJ Groenewegen, HW Berendse, *et al.* (2010) Diagnostic cerebrospinal fluid biomarkers for Parkinson's disease: a pathogenetically based approach. *Neurobiol Dis* 39: 229–241.

89. J Zhang, TJ Montine (2007) Proteomic discovery of CSF biomarkers for Alzheimer's disease. *Ann Neurol* 61: 497; author reply 497–498.

90. A Hye, S Lynham, M Thambisetty, M Causevic, J Campbell, HL Byers, *et al.* (2006) Proteome-based plasma biomarkers for Alzheimer's disease. *Brain* 129: 3042–3050.

91. DA Butterfield, D Boyd-Kimball, A Castegna (2003) Proteomics in Alzheimer's disease: insights into potential mechanisms of neurodegeneration. *J Neurochem* 86: 1313–1327.

92. S Lovestone, A Güntert, A Hye, S Lynham, M Thambisetty, M Ward (2007) Proteomics of Alzheimer's disease: understanding mechanisms and seeking biomarkers. *Expert Rev Proteomics* 4: 227–238.

93. MA Korolainen, TA Nyman, T Aittokallio, T Pirttilä (2010) An update on clinical proteomics in Alzheimer's research. *J Neurochem* 112: 1386–1414.

94. L Jiang, K Lindpaintner, HF Li, NF Gu, H Langen, L He, *et al.* (2003) Proteomic analysis of the cerebrospinal fluid of patients with schizophrenia. *Amino Acids* 25: 49–57.

95. SE Lakhan (2003) Diversification of U.S. medical schools via affirmative action implementation. *BMC Med Educ* 3: 6.

96. J Ottervald, B Franzén, K Nilsson, LI Andersson, M Khademi, B Eriksson, *et al.* (2010) Multiple sclerosis: Identification and clinical evaluation of novel CSF biomarkers. *J Proteomics* 73: 1117–1132.

97. MP Stoop, LJ Dekker, MK Titulaer, PC Burgers, PA Sillevis Smitt, TM Luider, *et al.* (2008) Multiple sclerosis-related proteins identified in cerebrospinal fluid by advanced mass spectrometry. *Proteomics* 8: 1576–1585.

98. J Guo, Z Sun, S Xiao, D Liu, G Jin, E Wang, *et al.* (2009) Proteomic analysis of the cerebrospinal fluid of Parkinson's disease patients. *Cell Res* 19: 1401–1403.

99. DC German, P Gurnani, A Nandi, HR Garner, W Fisher, R Diaz-Arrastia, *et al.* (2007) Serum biomarkers for Alzheimer's disease: proteomic discovery. *Biomed Pharmacother* 61: 383–389.

100. J Zhang, DR Goodlett, JF Quinn, E Peskind, JA Kaye, Y Zhou, *et al.* (2005) Quantitative proteomics of cerebrospinal fluid from patients with Alzheimer disease. *J Alzheimers Dis* 7: 125–133; discussion 173–180.

101. J Zhang, DR Goodlett, TJ Montine (2005) Proteomic biomarker discovery in cerebrospinal fluid for neurodegenerative diseases. *J Alzheimers Dis* 8: 377–386.

102. H Zetterberg, U Rüetschi, E Portelius, G Brinkmalm, U Andreasson, K Blennow, *et al.* (2008) Clinical proteomics in neurodegenerative disorders. *Acta Neurol Scand* 118: 1–11.

103. AH Simonsen, J McGuire, VN Podust, H Davies, L Minthon, I Skoog, *et al.* (2008) Identification of a novel panel of cerebrospinal fluid biomarkers for Alzheimer's disease. *Neurobiol Aging* 29: 961–968.

104. AH Simonsen, J McGuire, O Hansson, H Zetterberg, VN Podust, HA Davies, *et al.* (2007) Novel panel of cerebrospinal fluid biomarkers for the prediction of progression to Alzheimer dementia in patients with mild cognitive impairment. *Arch Neurol* 64: 366–370.

105. M Puchades, SF Hansson, CL Nilsson, N Andreasen, K Blennow, P Davidsson (2003) Proteomic studies of potential cerebrospinal fluid protein markers for Alzheimer's disease. *Brain Res Mol Brain Res* 118: 140–146.

106. Y Hu, JP Malone, AM Fagan, RR Townsend, DM Holtzman (2005) Comparative proteomic analysis of intra- and interindividual variation in human cerebrospinal fluid. *Mol Cell Proteomics* 4: 2000–2009.

107. F Abdi, JF Quinn, J Jankovic, M McIntosh, JB Leverenz, E Peskind, *et al.* (2006) Detection of biomarkers with a multiplex quantitative proteomic platform in cerebrospinal fluid of patients with neurodegenerative disorders. *J Alzheimers Dis* 9: 293–348.

108. GN Yin, HW Lee, JY Cho, K Suk (2009) Neuronal pentraxin receptor in cerebrospinal fluid as a potential biomarker for neurodegenerative diseases. *Brain Res* 1265: 158–170.

109. R Sultana, HF Poon, J Cai, WM Pierce, M Merchant, JB Klein, *et al.* (2006) Identification of nitrated proteins in Alzheimer's disease brain using a redox proteomics approach. *Neurobiol Dis* 22: 76–87.

110. R Sultana, D Boyd-Kimball, HF Poon, J Cai, WM Pierce, JB Klein, *et al.* (2006) Redox proteomics identification of oxidized proteins in Alzheimer's disease hippocampus and cerebellum: an approach to understand pathological and biochemical alterations in AD. *Neurobiol Aging* 27: 1564–1576.

111. R Sultana, M Perluigi, DA Butterfield (2006) Redox proteomics identification of oxidatively modified proteins in Alzheimer's disease brain and *in vivo* and *in vitro* models of AD centered around Abeta(1-42). *J Chromatogr B Analyt Technol Biomed Life Sci* 833: 3–11.

112. E Tolosa, G Wenning, W Poewe (2006) The diagnosis of Parkinson's disease. *Lancet Neurol* 5: 75–86.

113. AJ Hughes, SE Daniel, AJ Lees (2001) Improved accuracy of clinical diagnosis of Lewy body Parkinson's disease. *Neurology* 57: 1497–1499.

114. AJ Hughes, SE Daniel, L Kilford, AJ Lees (1992) Accuracy of clinical diagnosis of idiopathic Parkinson's disease: a clinico-pathological study of 100 cases. *J Neurol Neurosurg Psychiatry* 55: 181–184.

115. MG Spillantini, ML Schmidt, VM Lee, JQ Trojanowski, R Jakes, M Goedert (1997) Alpha-synuclein in Lewy bodies. *Nature* 388: 839–840.

116. VN Uversky (2003) A protein-chameleon: conformational plasticity of alpha-synuclein, a disordered protein involved in neurodegenerative disorders. *J Biomol Struct Dyn* 21: 211–234.

117. OM El-Agnaf, SA Salem, KE Paleologou, LJ Cooper, NJ Fullwood, MJ Gibson, *et al.* (2003) Alpha-synuclein implicated in Parkinson's disease is present in extracellular biological fluids, including human plasma. *FASEB J* 17: 1945–1947.

118. BI Giasson, MS Forman, M Higuchi, LI Golbe, CL Graves, PT Kotzbauer, *et al.* (2003) Initiation and synergistic fibrillization of tau and alpha-synuclein. *Science* 300: 636–640.

119. GK Wenning, KA Jellinger (2005) The role of alpha-synuclein and tau in neurodegenerative movement disorders. *Curr Opin Neurol* 18: 357–362.

120. EJ Finehout, Z Franck, KH Lee (2005) Complement protein isoforms in CSF as possible biomarkers for neurodegenerative disease. *Dis Markers* 21: 93–101.

121. A Sinha, N Srivastava, S Singh, AK Singh, S Bhushan, R Shukla, *et al.* (2009) Identification of differentially displayed proteins in cerebrospinal fluid of Parkinson's disease patients: a proteomic approach. *Clin Chim Acta* 400: 14–20.

122. I Rite, S Argüelles, JL Venero, S García-Rodriguez, A Ayala, J Cano, *et al.* (2007) Proteomic identification of biomarkers in the cerebrospinal fluid in a rat model of nigrostriatal dopaminergic degeneration. *J Neurosci Res* 85: 3607–3618.

Chapter 7

Liver Disease-Related Proteomics

Hirofumi Uto,[a] Yoichiro Takami,[b] Shuji Kanmura,[a] and Hirohito Tsubouchi[a]

[a]*Digestive and Lifestyle Diseases, Department of Human and Environmental Sciences, Kagoshima University Graduate School of Medical and Dental Sciences, 8-35-1 Sakuragaoka, Kagoshima, Kagoshima 890-8544, Japan*

[b]*Center for Clinical and Health Sciences, School of Pharmacy, Shujitsu University, 1-6-1 Nishigawara, Naka-ku, Okayama 703-8516, Japan*

hirouto@m2.kufm.kagoshima-u.ac.jp

Hepatocellular carcinoma (HCC) is the fifth most common cancer; advanced hepatic fibrosis is a major risk factor for HCC. Non-alcoholic fatty liver disease (NAFLD) can also progress to cirrhosis and HCC. These liver diseases and their progression are major causes of morbidity and mortality; identification of easily measurable and disease-specific biomarkers is required for early diagnosis and avoidance of invasive diagnostic procedures. Early diagnosis of HCC and assessment of the stage of hepatic fibrosis or NAFLD can also contribute to more effective therapeutic interventions and an improved prognosis. Thus far, we have identified several candidate biomarkers for liver diseases, including HCC and NAFLD, using proteomic technologies. Recent advances in methods for mass

Quantitative Proteome Analysis: Methods and Applications
Edited by Kazuhiro Imai and Sam F. Y. Li

ISBN 978-981-4316-51-4 (Hardcover), 978-981-4316-98-9 (eBook)
www.panstanford.com

spectrometric analysis, including protein labeling and amino acid analysis, facilitate highly sensitive and exhaustive proteomic analysis of patient samples. In this chapter, we will present an overview of proteomic technologies, including data from our laboratory. Our discussion will focus on the differential analysis of protein expression with the goal of discovering biomarkers for liver disease.

7.1 Introduction

In recent years, the mechanism of various diseases has been significantly clarified by remarkable developments in molecular biology, e.g., the decoding of the human genome. The clinical application of these developments has advanced considerably. To date, a number of attempts have been made to use the information obtained from single nucleotide polymorphisms (SNPs) and gene expression analyses using DNA microarray technique for diagnosis and treatment of various diseases. However, most physiological functions are controlled by proteins; the early diagnosis of disease may be difficult from analysis of gene expression alone. Therefore, the most feasible technology for near-term clinical application is proteomics.

Proteomics is the term used for the exhaustive analysis of protein structure and function in an organ or organism. Proteomics is useful for the elucidation of the pathology of HCC and chronic liver diseases and for the discovery of disease markers. Serum and plasma can be readily used as clinical samples because they are obtained using less invasive methods. If a biomarker associated with the pathology, disease progression or efficacy of treatment is identified in serum or plasma, this knowledge can be easily applied to early or differential diagnosis. Recently, proteomics research has spread rapidly as mass spectrometry technology undergoes remarkable advances; these methodological advances, including protein labeling and amino acid analysis, facilitate highly sensitive and exhaustive proteomic analysis of patient samples. These advances in proteomics techniques have promoted the exploration of biomarkers for malignant tumors, including HCC, and for chronic liver diseases, including liver cirrhosis and NAFLD.

In this chapter, we provide an overview of recent findings in the proteomic analysis of liver diseases, including results from our laboratory.

7.2 Diagnostic Biomarkers for Liver Disease Used Currently in a Clinical Setting

Early and accurate prognostic assessment of patients with liver diseases is difficult but critically important in order to choose the optimum clinical pathway. In acute liver failure, the most widely accepted multi-variable prognostic models, including the King's College Hospital (KCH) criteria [1–3] and the model for end-stage liver disease (MELD) score [4–6], have been improved by the incorporation of other clinical and biochemical indices [7, 8]. In addition, single-variable prognostic models such as serum Gc-globulin [9, 10], arterial blood lactate [8, 11], serum phosphate [12], arterial blood ammonia [13, 14], serum alpha-fetoprotein (AFP) [15–17], and prothrombin time [18] also appear to be promising markers. These markers have been used in many hospitals but should be further assessed in future large-scale prospective studies.

NAFLD is the hepatic manifestation of metabolic syndrome and is one of the most common causes of chronic liver disease worldwide [19, 20]. NAFLD represents a spectrum that includes simple steatosis and non-alcoholic steatohepatitis (NASH) [21]. Recently, increasing evidence suggests that only patients with histologic NASH follow a progressive course [21–23]. Several non-invasive biomarkers for NASH and hepatic fibrosis have been developed [24–30], but liver biopsy remains the gold standard for diagnosing NASH. However, liver biopsy is not easily tolerated, is associated with small but definite risks, and its results can be obscured by sampling errors. Therefore, the development of non-invasive diagnostic biomarkers for more sensitive and specific diagnosis of NASH and NASH-related fibrosis is important for use in clinical practice [23–25, 31]. Current evidence suggests that the pathogenesis of NASH involves the dysregulation of cytokines, adipokines, insulin resistance, and markers of apoptosis [32]. On the other hand, the development of fibrosis involves collagen deposition and other fibrogenic pathways. Although the two pathways may have common components, they are, nonetheless, distinct. Therefore, it may not be possible to develop a single test that can both diagnose NASH and stage the extent of fibrosis.

Diagnostic methods for hepatocellular carcinoma (HCC) include imaging, such as abdominal ultrasonography and computed tomography (CT), and measurement of serum tumor markers. Des-gamma-carboxy prothrombin (DCP, also known as PIVKA-II), AFP and the core fucosylated glycoform of AFP, AFP-L3, are widely used clinically as serum tumor markers for HCC [33–36]. The glycosylation of proteins is cell type-specific, so the N-linked glycosylation of a protein reflects modifications that occurred in the cell from which it came [37]. The glycosylation of the same protein may differ when it is secreted from diseased tissue, malignant cells or normal cells [38]. AFP can be produced in many circumstances, including in association with other liver diseases, and is not present in all cases of HCC [39, 40]. In fact, the sensitivity of AFP or DCP for detecting early-stage HCC is only 30–60% [33–36]. Therefore, the use of AFP as a primary screen for HCC has been the subject of suspicion; it is clear that more sensitive serum biomarkers for HCC are required.

7.3 Quantitative Proteomic Techniques for Liver Diseases

The discovery of easily measurable and disease-specific biomarkers in the tissue, blood and urine is very important for determinations of disease progression and evaluation of drug efficacy and toxicity. One of the strategies for exploring disease-specific biomarkers consists of two steps: (i) the identification of disease-specific proteins/peptides using tissues and cells from the animal disease models and/or patients and (ii) searching the distribution of identified proteins/peptides and the association between them and the disease in question. For the efficient discovery of biomarkers, more quantitative and reproducible techniques are required. Therefore, differential analysis of protein expression is frequently used in clinical proteomics. Quantitative proteomic approaches can be separated into both labeling and labeling-free methods and the labeling methods can be further separated into gel-based and non-gel–based methods (Fig. 7.1). The most typical gel-based differential method is two-dimensional fluorescence difference gel electrophoresis (2D-DIGE). On the other hand, non-gel–based methods include some stable isotope-labeling methods including cleavable isotope-coded affinity tags (cICAT) [41], stable isotope labeling by amino acids in cell culture (SILAC) [42], 2-nitrobenzenesulfenyl (NBS)-labeling method [43] and protein quantitation using amine-reactive isobaric

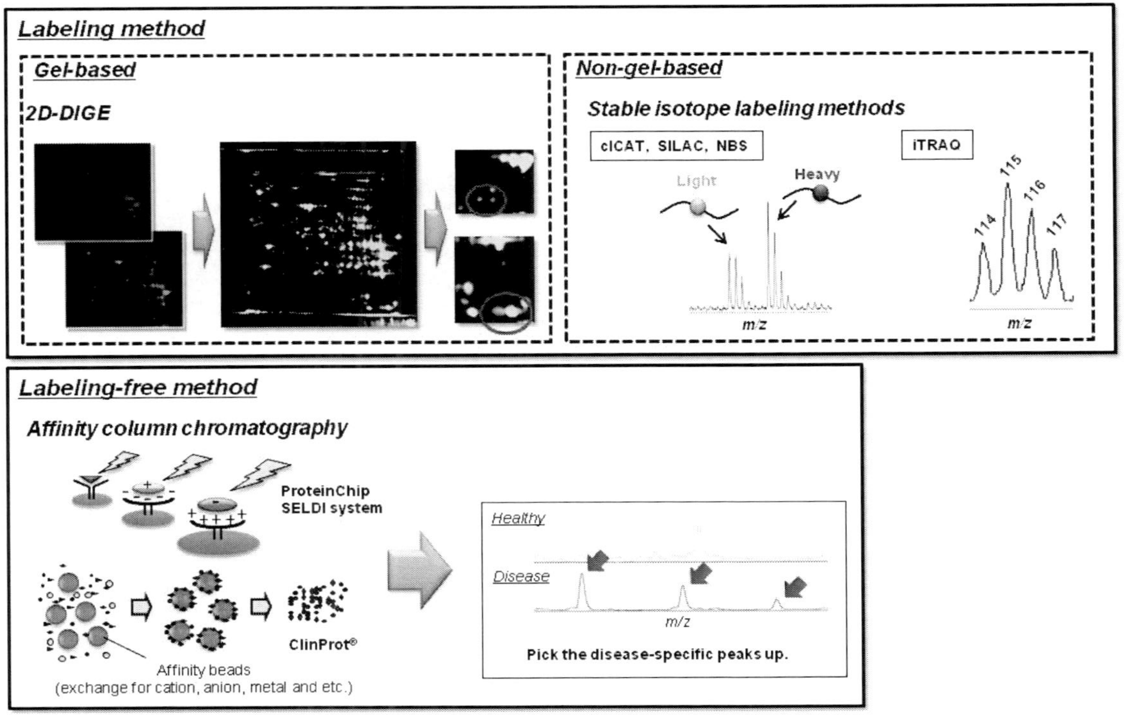

Figure 7.1 Schematic representation of quantitative proteomic methods. Quantitative proteomic approaches can be separated into labeling methods and labeling-free methods. Furthermore, labeling methods can be separated into gel-based and non-gel-based methods.

tagging reagents (iTRAQ) [44]. In addition, labeling-free methods such as surface-enhanced laser desorption ionization (SELDI) [45] and ClinProt® systems [46] based on affinity chromatography are useful for the analysis of blood samples (Fig. 7.1).

Here, we will introduce the methods for the differential analysis of protein expression and several recent findings regarding the identification of biomarkers for liver diseases.

7.3.1 2D-DIGE

The more classical gel-based methods include 2D-PAGE with proteins being separated in the first dimension by isoelectric point followed by separation in the second dimension by protein mass via sodium dodecyl sulfate polyacrylamide gel (SDS-PAGE). Technical experience is a significant factor in the successful application of 2D-PAGE, and a major obstacle is gel-to-gel variation due to running time and separation, which increases the difficulty in matching spots across gels. A newer technique is two-dimensional fluorescence difference gel electrophoresis (2D-DIGE). This method is conceptually similar to two-dye microarray experiments, in the sense that it employs cyanine (Cy) 2, Cy3, Cy5 or IC3-OSu, and IC5-OSu dyes [47] to enable the measurement of relative protein abundance within a single gel. Also, by allowing standardization across gels, the method enables the comparison of changes in protein expression between gels. Furthermore, the highly sensitive fluorescent dye, CyDye DIGE Fluor saturation dye, can be used [48, 49]. This dye, which labels cysteine residues, is a far more sensitive dye than silver staining; consequently, proteomic analyses using small number of cells or very small amounts of tissue have become possible [48, 49].

Several attempts to identify the novel biomarkers for liver fibrosis and HCC by 2D-DIGE have been reported so far. Qi *et al.* performed protein profiling of the diethylnitrosamine (DEN)-induced HCC tissue in rats using a 2D-DIGE-LC/MS system [50]. In their report, four subtypes of glutathione S-transferase (GSTP1, GSTP2, GSTM3, and GSTM5), the oxidoreductase-related protein [aldose reductase-like protein (ARLP), thioredoxin, peroxiredoxin 6 (PRX6)], and the 60 kDa heat shock protein (HSP60) were all upregulated in HCC tissue. Teramoto *et al.* compared protein expression between HCC samples and adjacent non-HCC samples in tissue from 18 HCC patients [51]. As mentioned above, 2D-DIGE is highly quantitative and reproducible and is therefore very useful for the elucidation of liver diseases using tissue samples. However, there are only a few reports of using serum/plasma samples for 2D-DIGE.

In 2D-PAGE of sera from patients with HCV-related chronic liver disease (CLD), the levels of inter-alpha-trypsin inhibitor heavy chain H4 (ITIH4) fragments, α1-antichymotrypsin, apolipoprotein L1, prealbumin, albumin, paraoxonase/arylesterase 1, and zinc-α2-glycoprotein were reduced, and those of CD5 antigen-like protein (CD5L) and β2 glycoprotein I (β2GPI) were elevated, in the liver cirrhosis group [52]. Using a similar method, the levels of Mac-2-binding protein, α2-macroglobulin and hemopexin were found to be elevated and those of α-1-antitrypsin, leucine-rich α2-glycoprotein and fetuin-A to be decreased in advanced liver fibrosis [53]. Ho *et al.* reported that the serum levels of α2-macroglobulin, vitamin D-binding protein and apolipoprotein AI, identified by 2D-DIGE, predicted liver fibrosis in hepatitis C patients [54]. However, in serum samples, albumin, globulin, transferrin, and antitrypsin account for about 90% of the serum protein composition, and the large amounts of these abundant proteins interfere with the separation of serum proteins by 2D-PAGE, making separation and analysis difficult. Thus, to analyze serum using 2D-PAGE or 2D-DIGE, the removal of these abundant proteins and subsequent detection of changes in trace protein levels are necessary. Ang *et al.* removed albumin from sera from patients with HCC and CLD by pretreating with lectin and then comparing the glycosylated haptoglobin expression level using 2D-PAGE [55]. The expression of glycosylated haptoglobin was increased in the HCC group, and the level was higher in advanced HCC compared with early-stage HCC, suggesting that glycated haptoglobin is useful for the diagnosis and prediction of HCC stage [55].

7.3.2 ProteinChip SELDI and ClinProt®

Surface-enhanced laser desorption ionization methods [45] and ClinProt® systems [46] based on affinity chromatography were useful for the analysis of blood samples (Fig. 7.1). The ProteinChip SELDI system (ProteinChip array®; Bio-Rad laboratories, Inc., CA, USA) has been successfully applied to diagnosis of various diseases. In this method, biological fluids such as serum and urine arc directly applied to spot surfaces without prior removal of abundant proteins.

7.3.2.1 ProteinChip SELDI

ProteinChip SELDI employs several types of spots whose surfaces are modified with various functional groups (hydrophobic, cationic, anionic, metal ion presenting and hydrophilic). After a series of binding and washing steps, protein mixtures retained on the surface are analyzed with mass spectrometry. Because differential analysis can be performed with a few micro-liters of serum or urine, ProteinChip SELDI is useful for identifying biomarkers in screens that use valuable specimens. However, the limitations of analysis include high cost and the fact that the MS used in the analysis is exclusive. To identify disease-specific proteins or peptides, MS/MS ion searching must be further employed.

The ProteinChip SELDI system allows two methods of serological diagnosis: one uses identification of individual proteins and functional analysis (Fig. 7.2A), whereas the other is based on a classification (decision tree) method established by data mining without protein identification (Fig. 7.2B). The identification of a protein corresponding to a target peak is difficult using the ProteinChip SELDI system; therefore, the classification-based diagnostic method (multi-marker analysis), in which identification of each protein corresponding to an individual peak is not necessary, is frequently used. The disease and control groups are differentiated based only on the expression levels of several protein peaks. We established a classification method based on six peaks (m/z 3444, 3890, 4067, 4435, 4470, and 7770) that were highly distinguishable between HCV-related HCC and HCV-related CLD and showed that this method is applicable for the diagnosis of both early-stage and advanced HCC [56]. This approach was capable of detecting HCC earlier than the detection of tumorous lesions by abdominal ultrasonography and was more useful for early diagnosis than current tumor markers such as AFP and DCP. Similarly, Zinkin *et al.* developed a diagnostic method using 11 protein peaks detected by three different types of protein chip used with the ProteinChip SELDI system. Their method had sensitivity and specificity for diagnosis of HCV related HCC of 79% and 86%, respectively; one of the 11 peaks, m/z 13391, was identified as the cystatin C [57]. The diagnostic sensitivity and specificity do not differ significantly from those of methods using current HCC markers (AFP, AFP-L3 fraction, and DCP), but the performance for the diagnosis of small HCC of ≤2 cm was better than that of methods using cur-

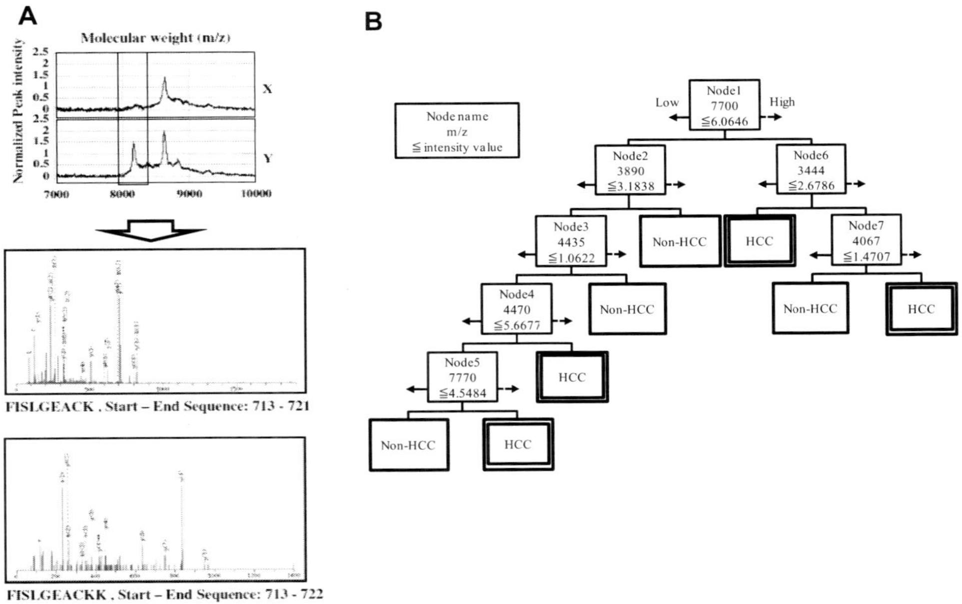

Figure 7.2 Serological proteomics using ProteinChip SELDI. (A) Identification of Complement 3a with sera from HCV-related HCC patients by ProteinChip SELDI and LC-MS/MS. (*J Gastroenterol*, Vol. 45, No. 4, 2010, 459–467. Reprinted with permission of Springer). (B) Classification of HCC and non-HCC samples via the decision tree. The decision tree was constructed to correctly classify 97% of the HCC samples in the first analysis group. The upper, middle, or lower lines in the box indicate the node name, molecular weight, and intensity value, respectively. (*Hepatology*, Vol. 45, No. 4, 2007, 948–956. Reprinted with permission of John Wiley and Sons).

rent markers. Recently, we found that five protein peaks (m/z 4067, 4470, 7564, 7929, and 8130) by ProteinChip SELDI were significantly increased in the sera of HCV-related HCC patients compared with HCV-related CLD patients and healthy subjects. Furthermore, one of these protein peaks, m/z 8130, was identified as the complement component 3a (C3a) fragment (Fig. 7.2A) [58].

7.3.2.2 ClinProt®

ClinProt® (Bruker Daltonics, Bremen, Germany) is a purification method based on affinity beads [46]. This method uses an outer layer of magnetic beads with chromatographic surfaces (hydrophobic, cationic, anionic, and metal ion presenting). As with ProteinChip SELDI, serum and urine samples are directly applied to the ClinProt® system without prior removal of abundant proteins. Proteins bound to the magnetic beads are then eluted and directly analyzed by mass spectrometry. The technical performance of affinity bead purification is similar to that of ELISA, and it can be used to process many samples in parallel. Several serum biomarkers for muscular dystrophy, type 1 diabetic nephropathy and renal cell carcinoma have been reported [59–61]. However, the reports of identification of useful biomarkers are still few, especially in the area of liver diseases.

7.3.3 Isotope Labeling of Proteins with SILAC, ICAT, and NBS Methods

Stable isotope-labeling methods include cICAT, SILAC, and NBS labeling. These methods utilize a light (^{12}C- or ^{14}N-labeled) and heavy (^{13}C- or ^{15}N-labeled) isotope to label different protein mixtures, allowing differential protein expression to be quantitatively detected in the same mass spectra.

7.3.3.1 Cleavable isotope-coded affinity tags

Cleavable isotope-coded affinity tags is one of the most frequently used isotope-labeling methods. Free cysteine thiol groups in peptides are targeted with the iodoacetamide-based ICAT reagent (Applied Biosystems, Inc., Foster City, CA) under strongly reducing conditions, with either a heavy or light isotope tag. An advantage of the cICAT reagents is that after digestion and mixing of the labeled peptides, one can further separate the mixture through a streptavidin column because each cICAT label contains a biotin covalently linked to an ethylene glycol linker group. This feature further reduces the complexity of the protein mixture before peptide identification and removes non-labeled peptides from the experimental workflow. cICAT labeled peptide pairs are then identified by MS/MS ion searching. A limitation of cICAT is that the peptide must contain a cysteine in order for labeling to occur. Therefore, ICAT may label only relatively few peptides of a given protein, leading to fewer peptides available for quantitation. cICAT is often combined with 2D-PAGE and LC-MS/MS techniques to improve the efficacy of identification of proteins.

Li *et al.* coupled laser capture microdissection with cICAT and two-dimensional liquid chromatography tandem mass spectrometry (2D-LC-MS/MS), and identified 261 proteins that are site-specifically increased or decreased in HCC [62]. In analysis using cICAT and LC-electrospray ionization tandem mass spectrometry (LC-ESI-MS/MS), Kang *et al.* compared serum proteins between nine cases of HCC and nine cases of hepatic cirrhosis and identified 31 proteins with differences in expression levels. Of these proteins, significantly enhanced expression of alpha-1-acid glycoprotein (AGP) was observed in the HCC validation group (patients with HCC or liver cirrhosis), showing that AGP is a candidate serum diagnostic marker for HCC [63]. Thus, serum proteomic analysis using a combination of cICAT and LC-ESI-MS/MS can be used for direct identification of potential protein biomarkers.

7.3.3.2 Stable isotope labeling with amino acids in cell culture

Stable isotope labeling with amino acids in cell culture incorporates a light or heavy labeled amino acid into two different cell populations growing in cell culture media (e.g., ^{12}C- and ^{13}C-labeled L-lysine or arginine) [64–67]. As cells grow and divide, the labeled amino acid is incorporated into proteins without compromising protein function. The cell cultures can then be used to perform proteomic profiling. In contrast, serum, tissue and other *in vivo* samples are not suitable for the SILAC.

Using the SILAC technique, several target proteins in HCC were identified. Chen *et al.* compared proteins expressed in MHCC97L with HCCLM6 (respectively, weakly and strongly metastatic HCC cell lines) and identified solute carrier family 12 member 2 (SLC12A2) and protein disulfide-isomerase A4 (PDIA4) as highly expressed in HCCLM6 [68]. Moreover, serum levels of both SLC12A2 and PDIA4 were significantly higher in strongly metastatic sera than in weakly metastatic sera derived from HCC patients [68]. Ren *et al.* compared the proteins expressed in the immortal hepatic cell line, L02, with those of the HepG2 hepatoma cell line; they identified phosphoglycerate mutase 1 (PGAM1) as highly expressed in HepG2. Furthermore, PGAM1 expression in HCC tissue was strongly correlated with poor differentiation and decreased survival rates [69]. Thus, proteomic research using cultured cell lines can discover candidates for clinical targets in HCC.

7.3.3.3 NBS labeling method

Like cICAT, NBS labeling method is a stable isotope-labeling method. This method was developed by Kuyama *et al.* [43], and is based on the specific binding reaction of the NBS reagent to tryptophan (W) residues within a protein; the 6-Da mass difference between [^{12}C]-NBS–labeled and [^{13}C]-NBS–labeled peptides generates a mass signature for all tryptophan-containing peptides. Several arylsulfenyl halides have selective reactivity towards the indole ring of tryptophan under acidic conditions [70]. Thus, 2-nitrobenezenesulfenyl chloride (NBS-Cl) effectively labels tryptophan residues. In this method, two protein samples are labeled with light and heavy NBS reagents, leading to a mass difference of 6 Da between the samples for all of the tryptophan-containing peptides. The labeled peptides are enriched by taking advantage of the relatively stronger affinity of NBS-labeled tryptophan-containing peptides for Sephadex media (LH-20) [71]. Relative quantitation of the proteins in the two samples is calculated from the intensities of paired peaks with a 6-Da mass difference in the MS spectra; proteins are then identified by queries based on data from the MS/MS ion searching (Fig. 7.3). Because tryptophan is the least abundant amino acid in proteins [72], isolation of tryptophan-labeled peptides reduces the number of samples and the complexity of the entire analysis. Most proteins (>90%) in *Homo sapiens* contain at least one tryptophan residue [72], so this method is suitable for global proteome analysis. However, one limitation of this method is that it is not suitable to peptidomic application, because the coverage of tryptophan labeling becomes smaller as the sequences of targets become shorter and shorter.

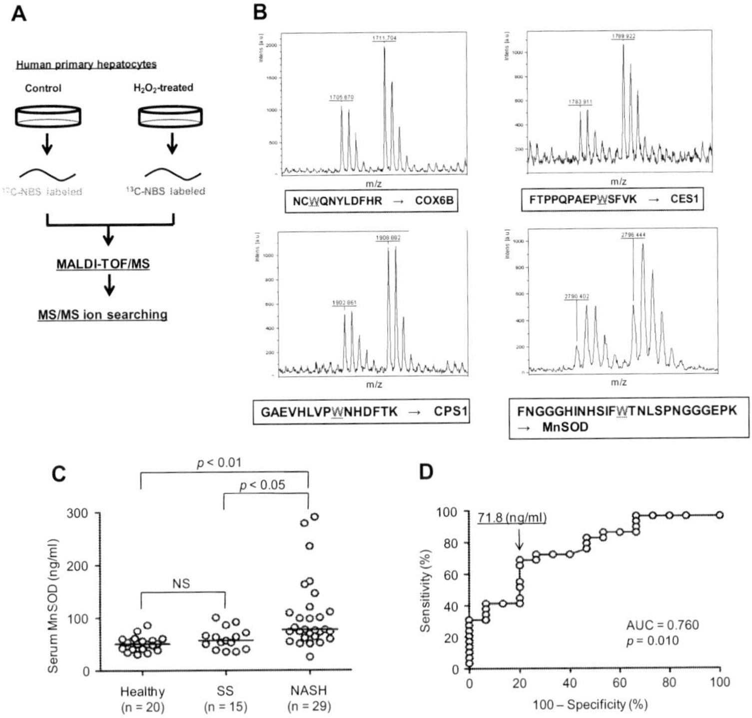

Figure 7.3 (A) Schematic representation of identification of oxidative stress markers in primary human hepatocytes using the 2-nitrobenzenesulfenyl (NBS) labeling method. (B) Four paired peaks with 6-Da differences and identification by MS/MS. (C) Serum MnSOD levels in healthy subjects and patients with simple steatosis (SS) or non-alcoholic steatohepatitis (NASH). (D) Receiver–operator curve for MnSOD to distinguish NASH (29 patients) from SS (15 patients) and a value of area under curve (AUC). (*Hepatol Res*, Vol. 40, No. 4, 2010, 438–445. Reprinted with permission of John Wiley and Sons).

We had previously reported the novel oxidative stress marker candidates that were useful for the diagnosis of the progression of NASH and HCV-related liver diseases. We attempted the NBS labeling method to identify novel oxidative stress markers, and identified four candidates: cytochrome *c* oxidase VIb isoform 1 (COX6B), liver carboxylesterase 1 (CES1), carbamoyl-phosphate synthase [ammonia], mitochondrial 1 (CPS1), and superoxide dismutase [manganese (Mn)], mitochondrial (MnSOD) in human primary hepatocytes (Fig. 7.3A,B) [30]. Furthermore, we found that NASH patients had significantly higher serum MnSOD levels than both healthy subjects and patients with simple steatosis (SS) (Fig. 7.3C,D) [30]. Another identified protein, CES1, may also be a diagnostic biomarker; Na *et al.* analyzed plasma by nano-LC/MS-MS and demonstrated increased expression of human liver carboxylesterase 1 (hCES1) in HCC patients [73].

7.3.3.4 Isobaric tagging for relative and absolute quantitation (iTRAQ)

Recently, isobaric tagging for relative and absolute quantitation of proteins (iTRAQ, Applied Biosystems) has been developed [44]. The functional components of the isobaric tagging system include a reporter tag (114–117 m/z), a balancer to ensure labeled peptides from different treatments have identical mass, and a peptide reactive group that chemically tags amine groups of peptides generated from tryptic digests. More recently, an eight-plex system has been developed with reporter tags 113–119 and 121 m/z to quantify protein expression. The fragmentation of the peptide tag generates low-molecular-mass reporter ions and high-molecular-mass cleaved fragment ions by MS/MS. Advantages of this method include increased ability to multiplex in a single LC-MS/MS experiment, and increased coverage of the proteome (because all peptides are theoretically labeled, in contrast to ICAT, which requires the presence of cysteine groups). In addition, multiple peptides can be quantified for a single protein, and confidence intervals can be calculated to obtain statistical measures of protein changes.

Chaerkady *et al.* compared the expressed proteins in HCC with the adjacent normal liver tissue and identified 59 proteins, including fibroleukin and myeloid associated differentiation marker, that are overexpressed in HCC tissue [74]. A combination of iTRAQ and 2D-LC-MS/MS has also been developed [75, 76]. Using the combined method, Chen's group compared the differences in protein expression between HepG2 hepatoma cells with hepatitis B virus (HBV)-infected HepG2 cells to identify several downregulated proteins in HBV-infected cells, including S100A6 and annexin A2 [75, 76].

7.4. Conclusion

In this chapter, we introduce several methods for differential proteomic analyses for liver diseases. The differential analyses can be useful tools for discovering novel biomarkers. When serum and tissue are used as a source of sample, several problems still remain; these include the removal of abundant protein and reproducibility between experiments. However, it is possible to discover the useful biomarkers in clinical practice, even using cultured cells. In addition, methods that include LC-MS/MS rapidly improved the efficacy of protein identification. In the future, we expect that the development of such combined techniques will lead to the discovery of useful biomarkers and the elucidation of pathogenesis of liver diseases.

References

1. Cholongitas EB, Betrossian A, Leandro G, Shaw S, Patch D, Burroughs AK. King's criteria, APACHE II, and SOFA scores in acute liver failure. *Hepatology* 2006; 43: 882.
2. Anand AC, Nightingale P, Neuberger JM. Early indicators of prognosis in fulminant hepatic failure: an assessment of the King's criteria. *J Hepatol* 1997; 26: 62–68.
3. O'Grady JG, Alexander GJ, Hayllar KM, Williams R. Early indicators of prognosis in fulminant hepatic failure. *Gastroenterology* 1989; 97: 439–445.
4. Dhiman RK, Jain S, Maheshwari U, Bhalla A, Sharma N, Ahluwalia J, Duseja A, Chawla Y. Early indicators of prognosis in fulminant hepatic failure: an assessment of the model for end-stage liver disease (MELD) and King's College Hospital criteria. *Liver Transpl* 2007; 13: 814–821.
5. Schmidt LE, Larsen FS. MELD score as a predictor of liver failure and death in patients with acetaminophen-induced liver injury. *Hepatology* 2007; 45: 789–796.
6. Yantorno SE, Kremers WK, Ruf AE, Trentadue JJ, Podesta LG, Villamil FG. MELD is superior to King's college and Clichy's criteria to assess prognosis in fulminant hepatic failure. *Liver Transpl* 2007; 13: 822–828.
7. Biggins SW, Kim WR, Terrault NA, Saab S, Balan V, Schiano T, Benson J, Therneau T, Kremers W, Wiesner R, Kamath P, Klintmalm G. Evidence-based incorporation of serum sodium concentration into MELD. *Gastroenterology* 2006; 130: 1652–1660.
8. Bernal W, Donaldson N, Wyncoll D, Wendon J. Blood lactate as an early predictor of outcome in paracetamol-induced acute liver failure: a cohort study. *Lancet* 2002; 359: 558–563.

9. Schiødt FV, Bangert K, Shakil AO, McCashland T, Murray N, Hay JE, Lee WM; Acute Liver Failure Study Group. Predictive value of actin-free Gc-globulin in acute liver failure. *Liver Transpl* 2007; 13: 1324–1329.
10. Antoniades CG, Berry PA, Bruce M, Cross TJ, Portal AJ, Hussain MJ, Bernal W, Wendon JA, Vergani D. Actin-free Gc globulin: a rapidly assessed biomarker of organ dysfunction in acute liver failure and cirrhosis. *Liver Transpl* 2007; 13: 1254–1261.
11. Schmidt LE, Larsen FS. Prognostic implications of hyperlactatemia, multiple organ failure, and systemic inflammatory response syndrome in patients with acetaminophen-induced acute liver failure. *Crit Care Med* 2006; 34: 337–343.
12. Baquerizo A, Anselmo D, Shackleton C, Chen TW, Cao C, Weaver M, Gornbein J, Geevarghese S, Nissen N, Farmer D, Demetriou A, Busuttil RW. Phosphorus ans an early predictive factor in patients with acute liver failure. *Transplantation* 2003; 75: 2007–2014.
13. Bhatia V, Singh R, Acharya SK. Predictive value of arterial ammonia for complications and outcome in acute liver failure. *Gut* 2006; 55: 98–104.
14. Bernal W, Hall C, Karvellas CJ, Auzinger G, Sizer E, Wendon J. Arterial ammonia and clinical risk factors for encephalopathy and intracranial hypertension in acute liver failure. *Hepatology* 2007; 46: 1844–1852.
15. Murray-Lyon IM, Orr AH, Gazzard B, Kohn J, Williams R. Prognostic value of serum alpha-fetoprotein in fulminant hepatic failure including patients treated by charcoal haemoperfusion. *Gut* 1976; 17: 576–580.
16. Schmidt LE, Dalhoff K. Alpha-fetoprotein is a predictor of outcome in acetaminophen-induced liver injury. *Hepatology* 2005; 41: 26–31.
17. Yang SS, Cheng KS, Lai YC, Wu CH, Chen TK, Lee CL, Chen DS. Decreasing serum alpha-fetoprotein levels in predicting poor prognosis of acute hepatic failure in patients with chronic hepatitis B. *J Gastroenterol* 2002; 37: 626–632.
18. Harrison PM, O'Grady JG, Keays RT, Alexander GJ, Williams R. Serial prothrombin time as prognostic indicator in paracetamol induced fulminant hepatic failure. *BMJ* 1990; 301: 964–966.
19. Marchesini G, Bugianesi E, Forlani G, Cerrelli F, Lenzi M, Manini R, Natale S, Vanni E, Villanova N, Melchionda N, Rizzetto M. Nonalcoholic fatty liver, steatohepatitis, and the metabolic syndrome. *Hepatology* 2003; 37: 917–923.
20. Younossi ZM, Diehl AM, Ong JP. Nonalcoholic fatty liver disease: an agenda for clinical research. *Hepatology* 2002; 35: 746–752.

21. Matteoni CA, Younossi ZM, Gramlich T, Boparai N, Liu YC, McCullough AJ. A non-alcoholic fatty liver disease: a spectrum of clinical and pathologic severity. *Gastroenterology* 1999; 116: 1413–1419.

22. Rafiq N, Bai C, Fang Y, Srishord M, McCullough A, Gramlich T, Younossi ZM. Long-term follow-up of patients with non-alcoholic fatty liver. *Clin Gastro Hepatol* 2009; 7: 234–238.

23. Ekstedt M, Franzén LE, Mathiesen UL, Thorelius L, Holmqvist M, Bodemar G, Kechagias S. Long-term follow-up of patients with NAFLD and elevated liver enzymes. *Hepatology* 2006; 44: 865.

24. Rosenberg WM, Voelker M, Thiel R, Becka M, Burt A, Schuppan D, Hubscher S, Roskams T, Pinzani M, Arthur MJ; European Liver Fibrosis Group. Serum markers detect the presence of liver fibrosis: a cohort study. *Gastroenterology* 2004; 127: 1704–1713.

25. Guha IN, Parkes J, Roderick P, Chattopadhyay D, Cross R, Harris S, Kaye P, Burt AD, Ryder SD, Aithal GP, Day CP, Rosenberg WM. Noninvasive markers of fibrosis in nonalcoholic fatty liver disease: validating the European liver fibrosis panel and exploring simple markers. *Hepatology* 2008; 47: 455–460.

26. Wieckowska A, Zein NN, Yerian LM, Lopez AR, McCullough AJ, Feldstein AE. *In vivo* assessment of liver cell apoptosis as a novel biomarker of disease severity in nonalcoholic fatty liver disease. *Hepatology* 2006; 44: 27–33.

27. Baranova A, Younossi ZM. The future is around the corner: noninvasive diagnosis of the progressive nonalcoholic steatohepatitis. *Hepatology* 2008; 47: 373–375.

28. Hossain N, Afendy A, Stepanova M, Nader F, Srishord M, Goodman Z, Younossi Z. Independent predictors of fibrosis in patients with non-alcoholic fatty liver disease. *Clin Gastro Hepatol* 2009; 7: 1224–1229.

29. Sumida Y, Nakashima T, Yoh T, Furutani M, Hirohama A, Kakisaka Y, Nakajima Y, Ishikawa H, Mitsuyoshi H, Okanoue T, Kashima K, Nakamura H, Yodoi J. Serum thioredoxin levels as a predictor of steatohepatitis in patients with nonalcoholic fatty liver disease. *J Hepatol* 2003; 38: 32–38.

30. Takami Y, Uto H, Tamai T, Sato Y, Ishida YI, Morinaga H, Sakakibara Y, Moriuchi A, Oketani M, Ido A, Nakajima T, Okanoue T, Tsubouchi H. Identification of a novel biomarker for oxidative stress induced by hydrogen peroxide in primary human hepatocytes using the 2-nitrobenzenesulfenyl chloride isotope labeling method. *Hepatol Res* 2010; 40: 438–445.

31. Miele L, Forgione A, Gasbarrini G, Grieco A. Noninvasive assessment of fibrosis in non-alcoholic fatty liver disease (NAFLD) and non-alcoholic steatohepatitis (NASH). *Transl Res* 2007; 149: 114–125.

32. Chitturi S, Farell GC. Etiopathogenesis of nonalcoholic steatohepatitis. *Semin Liver Dis* 2001; 21: 27–41.

33. Oka H, Tamori A, Kuroki T, Kobayashi K, Yamamoto S. Prospective study of alpha-fetoprotein in cirrhotic patients monitored for development of hepatocellular carcinoma. *Hepatology* 1994; 19: 61–66.

34. Ishii M, Gama H, Chida N, Ueno Y, Shinzawa H, Takagi T, Toyota T, Takahashi T, Kasukawa R. Simultaneous measurements of serum alpha-fetoprotein and protein induced by vitamin K absence for detecting hepatocellular carcinoma. South Tohoku District Study Group. *Am. J. Gastroenterol* 2000; 95: 1036–1040.

35. Okuda H, Nakanishi T, Takatsu K, Saito A, Hayashi N, Takasaki K, Takenami K, Yamamoto M, Nakano M. Serum levels of des-gamma-carboxyprothrombin measured using the revised enzyme immunoassay kit with increased sensitivity in relation to clinicopathologic features of solitary hepatocellular carcinoma. *Cancer* 2000; 88: 544–549.

36. Marrero JA, Su GL, Wei W, Emick D, Conjeevaram HS, Fontana RJ, Lok AS. Des-gamma carboxyprothrombin can differentiate hepatocellular carcinoma from nonmalignant chronic liver disease in American patients. *Hepatology* 2003; 37: 1114–1121.

37. Rudd PM, Dwek RA. Glycosylation: heterogeneity and the 3D structure of proteins. *Crit Rev Biochem Mol Biol* 1997; 32: 1–100.

38. Aoyagi Y, Suzuki Y, Igarashi K, Yokota T, Mori S, Suda T, Naitoh A, Isemura M, Asakura H. Highly enhanced fucosylation of alpha-fetoprotein in patients with germ cell tumor. *Cancer* 1993; 72: 615–618.

39. Ruoslahti E, Salaspuro M, Pihko H, Andersson L, Seppala M. Serum alpha-fetoprotein: diagnostic significance in liver disease. *Br Med J* 1974; 2: 527–529.

40. Di Bisceglie AM, Hoofnagle JH. Elevations in serum alpha-fetoprotein levels in patients with chronic hepatitis B. *Cancer* 1989; 64: 2117–2120.

41. Gygi SP, Rist B, Gerber SA, Turecek F, Gelb MH, Aebersold R. Quantitative analysis of complex protein mixtures using isotope-coded affinity tags. *Nat. Biotechnol* 1999; 17: 994–999.

42. Everley PA, Krijgsveld J, Zetter BR, Gygi SP. Quantitative cancer proteomics: stable isotope labeling with amino acids in cell culture (SILAC) as a tool for prostate cancer research. *Mol. Cell Proteomics* 2004; 3: 729–735.

43. Kuyama H, Watanabe M, Toda C, Ando E, Tanaka K, Nishimura O. An approach to quantitative proteome analysis by labeling tryptophan residues. *Rapid Commun Mass Spectrom* 2003; 17: 1642–1650.

44. DeSouza L, Diehl G, Rodrigues MJ, Guo J, Romaschin AD, Colgan TJ, Siu KW. Search for cancer markers from endometrial tissues using differentially labeled tags iTRAQ and cICAT with multidimensional liquid chromatography and tandem mass spectrometry. *J. Proteome Res* 2005; 4: 377–386.

45. Merchant M, Weinberger SR. Recent advancements in surface-enhanced laser desorption/ionization-time of flight-mass spectrometry. *Electrophoresis* 2000; 21: 1164–1177.

46. Cheng AJ, Chen LC, Chien KY, Chen YJ, Chang JT, Wang HM, Liao CT, Chen IH. Oral cancer plasma tumor marker identified with bead-based affinity-fractionated proteomic technology. *Clin. Chem* 2005; 51: 2236–2244.

47. Takeshita M, Ishida YI, Akamatsu E, Ohmori Y, Sudoh M, Uto H, Tsubouchi H, Kataoka H. Proanthocyanidin from blueberry leaves suppresses expression of subgenomic hepatitis C virus RNA. *J Biol Chem* 2009; 284: 21165–21176.

48. Fujii K, Kondo T, Yokoo H, Okano T, Yamada M, Yamada T, Iwatsuki K, Hirohashi S. Database of two-dimensional polyacrylamide gel electrophoresis of proteins labeled with CyDye DIGE Fluor saturation dye. *Proteomics* 2006; 6: 1640–1653.

49. Kondo T, Hirohashi S. Application of highly sensitive dyes (CyDye DIGE Fluor saturation dyes) to laser microdissection and two-dimensional difference gel electrophoresis (2D-DIGE) for cancer proteomics. *Nat Protoc* 2006; 1: 2940–2956.

50. Qi Y, Chen X, Chan CY, Li D, Yuan C, Yu F, Lin MC, Yew DT, Kung HF, Lai L. Two-dimensional differential gel electrophoresis/analysis of diethylnitrosamine induced rat hepatocellular carcinoma. *Int J Cancer* 2008; 122: 2682–2688.

51. Teramoto R, Minagawa H, Honda M, Miyazaki K, Tabuse Y, Kamijo KI, Ueda T, Kaneko S. Protein expression profile characteristic to hepatocellular carcinoma revealed by 2D-DIGE with supervised learning. *Biochim Biophys Acta* 2008; 1784: 764–772.

52. Gangadharan B, Antrobus R, Dwek RA, Zitzmann N. Novel serum biomarker candidates for liver fibrosis in hepatitis C patients. *Clin Chem* 2007; 53: 1792–1799.

53. Cheung KJ, Tilleman K, Deforce D, Colle I, Van Vlierberghe H. The HCV serum proteome: a search for fibrosis protein markers. *J Viral Hepat* 2009; 16: 418–429.

54. Ho AS, Cheng CC, Lee SC, Liu ML, Lee JY, Wang WM, Wang CC. Novel biomarkers predict liver fibrosis in hepatitis C patients: alpha 2 macroglobulin, vitamin D binding protein and apolipoprotein AI. *J Biomed Sci* 2010; 17: 58 doi: 10.1186/1423-0127-17-58.

55. Ang IL, Poon TC, Lai PB, Chan AT, Ngai SM, Hui AY, Johnson PJ, Sung JJ. Study of serum haptoglobin and its glycoforms in the diagnosis of hepatocellular carcinoma: a glycoproteomic approach. *J. Proteome Res* 2006; 5: 2691–2700.

56. Kanmura S, Uto H, Kusumoto K, Ishida Y, Hasuike S, Nagata K, Hayashi K, Ido A, Stuver SO, Tsubouchi H. Early diagnostic potential for hepatocellular carcinoma using the SELDI ProteinChip system. *Hepatology* 2007; 45: 948–956.

57. Zinkin NT, Grall F, Bhaskar K, Otu HH, Spentzos D, Kalmowitz B, Wells M, Guerrero M, Asara JM, Libermann TA, Afdhal NH. Serum proteomics and biomarkers in hepatocellular carcinoma and chronic liver disease. *Clin Cancer Res* 2008; 14: 470–477.

58. Kanmura S, Uto H, Sato Y, Kumagai K, Sasaki F, Moriuchi A, Oketani M, Ido A, Nagata K, Hayashi K, Stuver SO, Tsubouchi H. The complement component C3a fragment is a potential biomarker for hepatitis C virus-related hepatocellular carcinoma. *J Gastroenterol* 2010; 45: 459–467.

59. Alagaratnam S, Mertens BJ, Dalebout JC, Deelder AM, van Ommen GJ, den Dunnen JT, 't Hoen PA. Serum protein profiling in mice: identification of Factor XIIIa as a potential biomarker for muscular dystrophy. *Proteomics* 2008; 8: 1552–1563.

60. Gianazza E, Mainini V, Castoldi G, Chinello C, Zerbini G, Bianchi C, Galbusera C, Stella A, Mauri G, Zoppis I, Magni F, Kienle MG. Different expression of fibrinopeptide A and related fragments in serum of type 1 diabetic patients with nephropathy. *J Proteomics* 2010; 73: 593–601.

61. Chinello C, Gianazza E, Zoppis I, Mainini V, Galbusera C, Picozzi S, Rocco F, Galasso G, Bosari S, Ferrero S, Perego R, Raimondo F, Bianchi C, Pitto M, Signorini S, Brambilla P, Mocarelli P, Galli Kienle M, Magni F. Serum biomarkers of renal cell carcinoma assessed using a protein profiling approach based on ClinProt technique. *Urology* 2010; 75: 842–847.

62. Li C, Hong Y, Tan YX, Zhou H, Ai JH, Li SJ, Zhang L, Xia QC, Wu JR, Wang HY, Zeng R. Accurate qualitative and quantitative proteomic analysis of clinical hepatocellular carcinoma using laser capture microdissection coupled with isotope-coded affinity tag and two-dimensional liquid

chromatography mass spectrometry. *Mol Cell Proteomics* 2004; 3: 399–409.

63. Kang X, Sun L, Guo K, Shu H, Yao J, Qin X, Liu Y. Serum protein biomarkers screening in HCC patients with liver cirrhosis by ICAT-LC-MS/MS. *J Cancer Res Clin Oncol* 2010; 136: 1151–1159.

64. Ong SE, Blagoev B, Kratchmarova I, Kristensen DB, Steen H, Pandey A, Mann M. Stable isotope labeling by amino acids in cell culture, SILAC, as a simple and accurate approach to expression proteomics. *Mol Cell Proteomics* 2002; 1: 376–386.

65. Ong SE, Kratchmarova I, Mann M. Properties of ^{13}C-substituted arginine in stable isotope labeling by amino acids in cell culture (SILAC). *J Proteome Res* 2003; 2: 73–181.

66. Ong SE, Mann M. Identifying and quantifying *in vivo* methylation sites by heavy methyl SILAC. *Nat Methods* 2004; 1: 119–126.

67. Ong SE, Mann M. A practical recipe for stable isotope labeling by amino acids in cell culture (SILAC). *Nat Protoc* 2006; 1: 2650–2660.

68. Chen N, Sun W, Deng X, Hao Y, Chen X, Xing B, Jia W, Ma J, Wei H, Zhu Y, Qian X, Jiang Y, He F. Quantitative proteome analysis of HCC cell lines with different metastatic potentials by SILAC. *Proteomics* 2008; 8: 5108–5118.

69. Ren F, Wu H, Lei Y, Zhang H, Liu R, Zhao Y, Chen X, Zeng D, Tong A, Chen L, Wei Y, Huang C. Quantitative proteomics identification of phosphoglycerate mutase a as a novel therapeutic target in hepatocellular carcinoma. *Mol Cancer* 2010; 9: 81.

70. Fontana A, Scoffone E, Benassi CA, Sulfenyl halides as modifying reagents for polypeptides and proteins. II. Modification of cysteinyl residues. *Biochemistry* 1968; 7: 980.

71. Sasagawa T, Titani K, Walsh KA. Selective isolation of tryptophan-containing peptides by hydrophobicity modulation. *Anal Biochem* 1983; 134: 224.

72. Cagney G, Amiri S, Premawaradena T, Lindo M, Emili A. In silico proteome analysis to facilitate proteomics experiments using mass spectrometry. *Proteome Sci* 2003; 1: 5.

73. Na K, Lee EY, Lee HJ, Kim KY, Lee H, Jeong SK, Jeong AS, Cho SY, Kim SA, Song SY, Kim KS, Cho SW, Kim H, Paik YK. Human plasma carboxylesterase 1, a novel serologic biomarker candidate for hepatocellular carcinoma. *Proteomics* 2009; 9: 3989–3999.

74. Chaerkady R, Harsha HC, Nalli A, Gucek M, Vivekanandan P, Akhtar J, Cole RN, Simmers J, Schulick RD, Singh S, Torbenson M, Pandey A,

Thuluvath PJ. A quantitative proteomic approach for identification of potential biomarkers in hepatocellular carcinoma. *J Proteome Res* 2008; 7: 4289–4298.

75. Niu D, Sui J, Zhang J, Feng H, Chen WN. iTRAQ-coupled 2-D LC-MS/MS analysis of protein profile associated with HBV-modulated DNA methylation. *Proteomics* 2009; 9: 3856–3868.
76. Feng H, Li X, Niu D, Chen WN. Protein profile in HBx transfected cells: a comparative iTRAQ-coupled 2D LC-MS/MS analysis. *J Proteomics* 2010; 73: 1421–1432.

Chapter 8

Respiratory Disease-Related Proteome

Masahiro Seike and Akihiko Gemma
Department of Pulmonary Medicine and Oncology,
Graduate School of Medicine, Nippon Medical School,
1-1-5 Sendagi, Bunkyo-ku, Tokyo, 113-8603, Japan
mseike@nms.ac.jp

8.1 Introduction

The identification of disease-specific proteins by proteome analysis has been tried to date for malignant tumors such as lung cancer and diffuse lung diseases, including idiopathic pulmonary fibrosis (IPF), sarcoidosis, and pulmonary alveolar proteinosis. Combinations of two-dimensional polyacrylamide gel electrophoresis (2D-PAGE), high-performance liquid chromatography (HPLC), and surface-enhanced laser desorption/ionization (SELDI) with mass spectrometry (MS) (2D-PAGE/MS, LC/MS/MS, and SELDI-TOF-MS, respectively) have been employed for analysis. This chapter outlines approaches to identifying proteins specific to these pulmonary diseases.

Quantitative Proteome Analysis: Methods and Applications
Edited by Kazuhiro Imai and Sam F. Y. Li

ISBN 978-981-4316-51-4 (Hardcover), 978-981-4316-98-9 (eBook)
www.panstanford.com

8.2 Lung Cancer

A number of studies have applied proteome analysis to diagnose lung cancer and explore prognostic factors and lung cancer-related factors associated with the treatment target, and findings of great interest with clinical relevance have been reported. The two main objectives are (1) early detection of lung cancer and identification of biomarkers to predict prognosis and (2) the identification of sensitivity and resistance factors to drugs currently available and development of molecular targeted agents aimed at controlling cancer-related factors.

Drug therapy regimens for lung cancer have currently been selected on the basis of histology. The identification of drug sensitivity factors such as epidermal growth factor tyrosine kinase inhibitors (EGFR-TKI) and factors associated with adverse effects would facilitate the development of novel therapeutic applications. The identification of tumor-specific proteins and peptides is sought by analytical methods using samples from not only lung cancer cell lines, tumor tissue and serum but also pleural effusion and urine. Proteome analysis using these samples is outlined in this text.

8.2.1 Cell Lines

Analyses using lung cancer cell lines have been carried out most frequently. The methods of 2D-PAGE/MS and LC/MS/MS have been often employed, and there are a number of reports on the occurrence and metastasis of lung cancer. Although analysis using a cell line will not completely reflect the protein profile in tumor tissue, this approach has been employed worldwide because of the advantages of easy protein extraction, excellent data reproducibility, and possible repetition of experiments.

8.2.2 Tumor Tissue

Protein analysis using lung cancer tissue obtained by surgery and bronchoscopy has been tried in many studies. Chen *et al.* carried out proteome analysis of lung cancer by 2D-PAGE/MS using lung cancer tissue, and identified 20 kinds of protein associated with the prognosis for patients with lung adenocarcinoma. Among them, an increase in serum phosphoglycerate kinase 1 (PGK-1) levels was strongly associated with poor prognosis.[1] This is a representative proteome analytical method to examine correlations with patient prognosis and drug sensitivity.

A tumor lesion consists of a variety of factors, such as stroma and blood vessels, in addition to tumor tissue, and these factors have various functions and roles. Although proteome analysis using all components of tumor tissue, stroma, and blood vessels has been carried out in previous reports, due to technical limitations it is difficult to evaluate what biological features are specifically reflected in the changes in protein expression found by these analyses.

In particular, to evaluate the profile of protein expression in highly heterogeneous lung cancer tissue, strict sample collection by laser capture microdissection, employed for gene expression analysis, is desirable. After extracting proteins from only tumor tissue by laser microdissection, Seike *et al.* successfully used two-dimensional difference gel electrophoresis (2D-DIGE) in combination with highly sensitive fluorescent dyes to identify protein expression patterns that were able to categorize histological types of lung cancer.[2] The evaluation of the biological features in each lesion from the aspect of specific protein expression can eventually enable the prediction of cancer metastatic potentials, disease prognosis, and sensitivity to chemotherapy.

When samples are collected by these highly accurate methods and expression profiles are compared between primary tumor tissue and metastatic lesions and between high and low metastatic cells, coupled with the analysis of expression profiles of stromal cell components, the identification of molecules involved in metastatic mechanisms of lung cancer may be achieved.

8.2.3 Serum/Plasma

Advances have been made in proteome analysis using serum/plasma proteins. There is a limitation in the collection of analytical samples from tumor tissue, whereas analysis using serum/plasma is less invasive and thus greatly beneficial for patients and suitable for screening for early detection of lung cancer. Candidate biomarkers for early detection are (1) proteins secreted directly from tumor tissue and (2) proteins secondarily changeable by biologically responding to tumors.

Employing 2D-DIGE/MS using serum from 10 patients with non-small cell lung cancer and 10 healthy subjects, Hoagland *et al.* identified haptoglobin and its specific repair after translation as candidate biomarkers for non-small cell lung cancer.[3] However, because there are a number of high-concentration high-molecular-weight proteins in the plasma, such as albumin and globulin, it is difficult to detect protein and peptides at the very low levels derived from tumor tissue.

Recently, biomarkers have been explored by analytical methods such as 2D-PAGE/MS and LC/MS/MS using samples after removal of albumin and globulin with affinity columns. These methods, which provide 10–20 times as many protein spots as analyses using whole serum, have enabled quantitative and comprehensive analysis.[4,5]

Today, the most appropriate analytical method for proteome analysis using serum/plasma is the LC-MS system. Detection of protein molecules at femtomolar levels has been possible in recent analyses. Ueda *et al.* reported that apolipoprotein-4, a protein degradation product specific to lung adenocarcinoma, was detected by LC-MS/MS using serum from 62 patients with lung adenocarcinoma.[6]

Proteome analysis targeted at serum/plasma protein plays a great role in clinical practice, and the advancement of analytical technologies has almost enabled the identification of targeted lung cancer-specific protein molecules. Further development of analytical methods is anticipated.

8.2.4 Pleural Effusion and Urine

For the exploration of biomarkers for early detection of lung cancer, analyses using pleural effusion and urine samples have been advanced. Tyan *et al.* identified 161 proteins by protein analysis by 2D-PAGE/MS using pleural effusion samples from patients with lung adenocarcinoma. It is intriguing that proteins specific to pleural effusion that had not been reported in serum protein studies were identified.[7]

Tantipaiboonwon *et al.* analyzed urine samples from lung cancer patients by 2D-PAGE/MS after ultrafiltration and reported that it was possible to identify cancer-related urine proteins.[8] Urine is the easiest sample to collect: Its collection is minimally invasive to patients, which provides great benefit. Further advancement of analytical methods in this area is expected.

8.3 Diffuse Lung Disease

Proteome studies have been carried out with samples from cell culture, tissue, serum, and bronchioalveolar lavage fluid (BALF), in addition to microsampled specimens from peripheral airways from patients with IPF, and pulmonary alveolar proteinosis. In particular, BALF, which reflects most accurately the protein composition in peripheral airways, has been used widely for diagnosis, and it provides critically important information in elucidating the pathogenesis of disease.

Proteome analysis by 2D-PAGE using BALF was done for the first time in 1979, and a database was constructed on the basis of the analysis of patients with pulmonary alveolar proteinosis and healthy non-smokers.[9] Since then, several studies have been reported mostly with regard to IPF and sarcoidosis. Lenz *et al.* analyzed BALF from patients with lung disease and healthy subjects and reported a difference in the expression of major proteins such as surfactant protein A (SP-A), IgG, and IGA.[10] Wattiez *et al.* identified proteins specific to IPF by 2D-PAGE using BALF.[11] Sabounchi-Schütt *et al.* identified 17 protein spots specific to sarcoidosis by 2D-PAGE using BALF.[12]

It is more invasive to collect BALF than serum, but BALF reflects the pathogenesis of lung disease more appropriately. Proteome analysis using BALF is possible with a small amount of sample due to the recent advancements of mass spectrometry, and it is expected to contribute to the detection of new biomarkers and the elucidation of disease mechanisms.

References

1. Chen G, Gharib TG, Wang H, *et al.* Protein profiles associated with survival in lung adenocarcinoma. *Proc Natl Acad Sci U S A* 2003; 100(23): 13537–13542.
2. Seike M, Kondo T, Fujii K, Okano T, Yamada T, Matsuno Y, Gemma A, Kudoh S, Hirohashi S. Proteomic signatures for histological types of lung cancer. *Proteomics* 2005; 5(11): 2939–2948.
3. Hoagland LF 4th, Campa MJ, Gottlin EB, Herndon JE 2nd, Patz EF, Jr. Haptoglobin and posttranslational glycan-modified derivatives as serum biomarkers for the diagnosis of nonsmall cell lung cancer. *Cancer* 2007 15; 110(10): 2260–2268.

4. Fujii K, Nakano T, Kanazawa M, Akimoto S, Hirano T, Kato H, Nishimura T. Clinical-scale high-throughput human plasma proteome analysis: lung adenocarcinoma. *Proteomics* 2005; 5(4): 1150–1159.
5. Okano T, Kondo T, Kakisaka T, Fujii K, Yamada M, Kato H, Nishimura T, Gemma A, Kudoh S, Hirohashi S. Plasma proteomics of lung cancer by a linkage of multi-dimensional liquid chromatography and two dimensional difference gel electrophoresis. *Proteomics* 2006; 6(13): 3938–3948.
6. Ueda K, Saichi N, Takami S, Kang D, Toyama A, Daigo Y, Ishikawa N, Kohno N, Tamura K, Shuin T, Nakayama M, Sato TA, Nakamura Y, Nakagawa H. A comprehensive peptidome profiling technology for the identification of early detection biomarkers for lung adenocarcinoma. *PLoS One* 2011; 6(4): 18567.
7. Tyan YC, Wu HY, Su WC, Chen PW, Liao PC. Proteomic analysis of human pleural effusion. *Proteomics* 2005; 5(4): 1062–1074.
8. Tantipaiboonwong P, Sinchaikul S, Sriyam S, Phutrakul S, Chen ST. Different techniques for urinary protein analysis of normal and lung cancer patients. *Proteomics* 2005; 5(4): 1140–1149.
9. Bell DY, Hook GE. Pulmonary alveolar proteinosis: analysis of airway and alveolar proteins. *Am Rev Respir Dis* 1979; 119(6): 979–990.
10. Lenz AG, Meyer B, Costabel U, Maier K. Bronchoalveolar lavage fluid proteins in human lung disease: analysis by two-dimensional electrophoresis. *Electrophoresis* 1993; 14(3): 242–244.
11. Wattiez R, Hermans C, Cruyt C, Bernard A, Falmagne P. Human bronchoalveolar lavage fluid protein two-dimensional database: study of interstitial lung diseases. *Electrophoresis* 2000; 21(13): 2703–2712.
12. Sabounchi-Schütt F, Aström J, Hellman U, Eklund A, Grunewald J. Changes in bronchoalveolar lavage fluid proteins in sarcoidosis: a proteomics approach. *Eur Respir J* 2003; 21(3): 414–420.

Chapter 9

Renal Disease-Related Proteome

Shinya Kaname[a] and Tadashi Yamamoto[b]
[a]*Department of Nephrology and Rheumatology, Kyorin University School of Medicine, 6-20-2, Shinkawa, Mitaka, Tokyo 181-8611, Japan*
[b]*Department of Structural Pathology, Institute of Nephrology, Graduate School of Medical and Dental Sciences, Niigata University, Niigata, Japan*
kaname@ks.kyorin-u.ac.jp

9.1 Overview of and Update on Proteomic Analysis for Renal Diseases

Proteomic analysis in nephrology has been intended mainly for the purpose of searching for diagnostic/predictive biomarkers or clarifying the pathogenetic mechanisms/factors in various renal diseases by using urine or kidney or sometimes plasma samples [1–3]. Human renal tissues utilized for proteomic analysis are mostly frozen samples of renal biopsies and sometimes formalin-fixed kidney tissues obtained by autopsy. Urine samples have been analyzed for finding new biomarkers by proteomics. Technical advances in methodology, especially for separating proteins from urine and renal tissues, along with a recent breakthrough of protein

Quantitative Proteome Analysis: Methods and Applications
Edited by Kazuhiro Imai and Sam F. Y. Li

ISBN 978-981-4316-51-4 (Hardcover), 978-981-4316-98-9 (eBook)
www.panstanford.com

identification techniques by mass spectrometry, have dramatically accelerated renal proteomic research, although there are not so many achievements so far, unfortunately, that can be applied into clinical practice.

In parallel to research progress, proteomic database of urinary and renal proteins have been constructed. The most comprehensive one is provided in the Web site of the Human Kidney and Urine Proteome Project (HKUPP), one of HUPO (Human Proteome Organization) initiatives, which confers open access worldwide via Web site (http://www.hkupp.org/) [4, 5]. Other large-scale databases of urine proteomes are MAPU (Max-Planck Unified) database [6], and proteome of rat renal collecting ducts (Table 9.1).

Table 9.1 Database of proteome for urine and kidney tissues

Name/ author	Sample	URL	References
HKUPP	Human urine and kidney tissues	http://hkupp.kir.jp/index.htm	4, 8, 9
EuroKUP	Human urine and kidney tissues	http://www.eurokup.org/	5
MAPU	Human urine	http://141.61.102.16/urine/	6, 36
Renal biobank catalogue	Human kidney tissues	http://www.p3gobservatory.org/network/studies.htm?filter=renal&partner=renal	
NIH	Rat IMCD proteome	http://dir.nhlbi.nih.gov/papers/lkem/imcd/	17
	Rat IMCS phosphoproteome	http://dir.nhlbi.nih.gov/papers/lkem/cdpd/	19
	human exosome	http://dir.nhlbi.nih.gov/papers/lkem/exosome/	42, 43
Urinebank (Japan)	Human urine	http://www.urinebank.org/	4

9.2 Samples and Methodologies of Proteomic Analysis for Renal Disease

9.2.1 Renal Tissues

The kidney has a complicated structure consisting of a number of different segments, each of which plays different roles orchestrating for the maintenance of body fluid and electrolyte balance. Functional unit is a nephron composed of glomerulus and the following tubular segments. In a glomerulus, where plasma filtration occurs, there are three cell types, including endothelial cells, epithelial cells (also called podocytes) and mesangeal cells, whereas the tubules are divided into functionally and anatomically distinct segments, namely proximal tubule, loop of Henle, distal tubule, and the cortical and medullary collecting duct, generating urine through reabsorption and secretion via tubular epithelia. Renal diseases are caused by functional or organic defect of various causes in any part or a whole of renal tissues, nephron segments, cells, or even molecule such as transporter in the tubules; thus, selective approach will be required to clarify the pathophysiology. In animal models, such a selective renal segment collection is feasible; thus, a number of basic proteomic research studies have been conducted in the last decade. In human analysis, the target tissues are mainly obtained by renal biopsy and sometimes by autopsy. Therefore, the major problem in human kidney proteome analysis may be that renal biopsy samples contain only small parts of the kidney yielding a tiny and usually insufficient amount of proteins [7]. However, recent progress in laser captured microdissection (LCM) and isolating techniques using FACS has made it possible to analyze the small sections or cells selectively from biopsy samples or selective tissues (see below). The sensitivities of protein identification have been improving but still not satisfactory, thus, more sensitive protein identification and analyzing techniques should be developed.

9.2.1.1 Normal kidney tissues

Normal renal tissues of animals or human are utilized for making proteome database to understand difference in profile of proteins in disease conditions and/or analyzing physiological functions in a whole kidney or a part of renal tissues. More intensive and selective approaches using isolated glomerulus or tubules without or with any treatment of choice are available in animal studies.

9.2.1.1.1 *Glomerulus*

Yoshida *et al.* were the first to perform comprehensive proteome analysis for normal human glomeruli in 2005, when they constructed two-dimensional maps of SDS-PAGE for glomerular proteins from the removed kidney by nephrectomy, followed by analysis of the spots using LC-MS and MALDI-TOF-MS. Total 347 protein spots representing 212 proteins had been identified in the study [8], their database being available on the HKUPP Web site. Later, Miyamoto *et al.* created the comprehensive database of normal human glomeruli and identified more than three thousands of proteins [9].

9.2.1.1.2 *Renal cortex/medulla*

Witzmann *et al.* applied mass spectrometry to the proteome analysis of rat renal cortex and medulla more than 10 years ago [10]. Subsequently, a couple of comprehensive studies have been reported: Sitek *et al.* used a two–dimensional DIGE (difference in gel electrophoresis) technique in a proteomic analysis of the glomeruli obtained by LCM (see Section 9.2.1.2.1) [11]. Zhao *et al.* conducted comprehensive proteomic analysis for renal cortex in mice using LC-MS, identifying two thousands of different proteins [12]. Zhang *et al.* constructed the phosphoproteome by examining tyrosine-phosphorylated proteins from isolated glomeruli in rats by two-dimensional PAGE followed by LC-MS/MS and identified slit diaphragm proteins such as nephrin, paxillin, and FAK1 (focal adhesion kinase 1) [13].

9.2.1.1.3 *Tubules*

Knepper's group has been conducting proteomic analysis for fractions of isolated rat inner medullary collecting duct (IMCD) cell lysates by LC-MS/MS, the database of which being available also on the HKUPP Web site [14–18]. They examined the proteins that were induced or phosphorylated by vasopressin stimulation, showing that hundreds of proteins were phosphorylated, including non-membrane proteins (myosin-4/10) and membrane proteins such as aquaporin-2 (AQP-2), urea transporters (UT-A1/3), chloride channel (ClC-1) and sodium-bicarbonate cotransporter (NBC3), all of which proved to have novel phosphorylation sites. It was also shown that AQP-2 was downregulated by water load, a mimic of vasopressin escape, and it was revealed, combined with bioinformatics of protein network, that the amounts of several other regulatory proteins with a low abundance were altered [15].

Nielsen *et al.* examined the proteome of rat IMCD by DIGE/MALDI-TOF MS, showing that lithium intake of the animal caused an increase in various signal transduction proteins, including MAPK and GSK3 beta, most of which may be related to a decrease in AQP2 [19].

Rivard *et al.* examined the effect of hyperosmolality on the protein profiles of rat IMCD, showing that S100-A4, a calcium-binding protein, was strongly induced [20].

Curthoys *et al.* analyzed proteome of proximal tubules isolated from kidneys of rats that were made metabolic acidosis by loading NH_4Cl. DIGE/MALDI-TOF-MS showed that 17 novel proteins were increased, in addition to already known ones such as glutaminase, glutamate dehydrogenase (GDH) and phosphoenolpyruvate carboxykinase (PEPCK). Moreover, the study showed that these slowly induced proteins had the AU-rich sites in common, which seemed to be associated with pH changes [21].

9.2.1.2 Disease kidney tissues (especially human kidney biopsy samples)

Renal biopsy samples have recently become available for proteomic research due to technical advances in isolation and analysis of a small amount of proteins. LCM is utilized for collecting glomerulus and other renal compartments selectively [22]. The findings obtained from the clinical or basic analysis for specific renal diseases will appear later separately.

9.2.1.2.1 *LCM from a technical aspect*

In 2005, Xu *et al.* first applied LCM to the proteomic analysis of the kidneys from model animal. Fifty glomerular sections were captured from frozen kidney tissues of 5/6 nephrectomized rats and the proteins were analyzed by MALDI-MS, showing that sclerotic and non-sclerotic glomeruli had distinct protein profiles [23]. Recently, formalin-fixed, paraffin-embedded (FFPE) tissues have become usable with almost comparable yields as those of frozen tissues by using commercially available kits such as Liquid Tissue (Expression Pathology), Qproteome (Qiagen) or FFPE Protein Extraction Solution (Agilent). In addition, new methods have been developed as the filter-aided sample preparation (FASP) [24] and the on-site Direct Digestion) methods (OSDD) [25]. Indeed, by applying the FASP method to the proteomic analysis of microdissected FFPE cancer tissues showed high retrieval of proteins, including known colon cancer markers and others [24]. The OSDD method can minimize the loss of peptides during the preparation, resulting in a high yield of proteins from microdissected sections. Sethi *et al.* first performed comparative proteomic analysis for glomeruli captured by LCM in the FFPE biopsy samples of the patients with type II membranoproliferative glomerulonephritis (MPGN type II) by using DIGE technique (see also Section 9.3.4) [26, 27].

Despite such a progress in LCM techniques, there remain several problems to be solved. More sensitive and quantitative techniques that can detect smaller amounts of proteins from the limited biopsy samples are clearly warranted. In the study by Sitek [11], the lysine residues of glomerular proteins were reacted with two different dyes such as C5 and C7, which were separated by two-dimensional gel electrophoresis, showing that less than 1 μg of proteins were sufficient to visualize protein spots. Moreover, this technique may allow protein semiquantification and also have other advantages that the protein charges that may affect the proteome profiles will not be changed during preparing FFPE samples because of the dyes being electrically neutral. Waanders *et al.* recently showed highly sensitive method of examining glomerular proteome by LC-MS and LCM-based isolation technique [28]. DeSouza *et al.* attempted semi-quantification for biomarkers using the microdissected FFPE samples by single/multiple reaction monitoring (SRM/MRM) method, in which mTRAQ (MRM tags for relative and absolute quantitation) labeling was combined with LC-MS/MS [29]. In the future, automated LCM approach that can deal with a large number of samples simultaneously will hopefully be investigated in a large-scale analysis.

9.2.2 Urine Proteomics

Urine is a sample that can be easily and noninvasively obtained with abundant information and thus has been thought to be an ideal material for proteome analysis of renal diseases [30–34]. Indeed, urine contains a variety of proteins deriving from the blood, kidney, and the urinary tract in soluble or insoluble forms, with their profiles and amounts varying depending on the extent of glomerular filtration and tubular absorption/secretion as well as the disease status of the kidney. Chen *et al.* performed one- or two-dimensional gel-electrophoresis to isolate urinary proteins in healthy people, followed by trypsinization of the excised bands in the gel and comprehensive mass spectrometry analysis, identifying more than a thousand of different proteins [35]. Thirty percent of them turned out to be derived from the renal tissues, whereas the other 30% was shown to be plasma or blood proteins such as albumin, hemoglobins, immunoglobulins, and other smaller proteins. Therefore, we should be careful that these common proteins may be increased more than 10-fold in glomerular diseases and hinder the analysis of the other target protein. Thus, in exploring biomarkers or proteome in the urine, the results vary depending upon the method of sample preparation, protein isolation and the disease state of the host (Fig. 9.1). It is also reported whether or not urinary sediments are contained in the sample may influence the result [36].

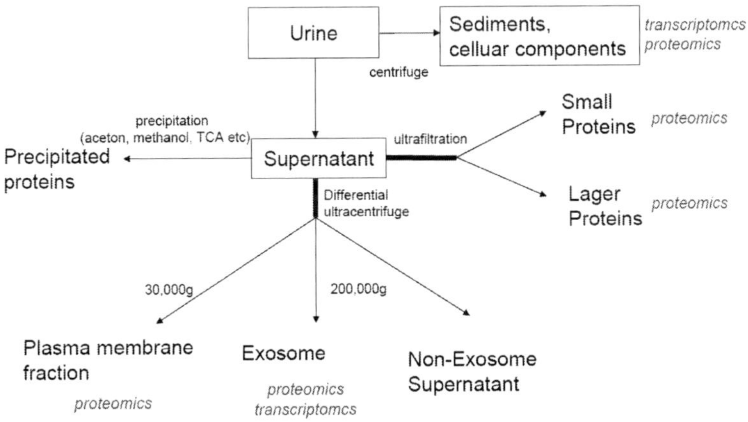

Figure 9.1 Method of Urine preparation for proteomic research

As described above, urine contains insoluble proteins in a form of various cells and cellular /tissue debris, casts, and membrane components or vesicles, including exosomes. Among them, urinary exosomes have been recently examined extensively and thus will be described later separately.

9.2.2.1 Database of urinary proteome

Adachi *et al.* performed a comprehensive proteome analysis of urine supernatant after simple centrifuge using LC-MS/MS in healthy people and identified more than 1500 proteins with nearly half belonging to membrane proteins, the lists of which are accessible on the Web [37]. Other large-scale databases for urine proteome are prepared by HKUPP and EuroKUP as described earlier (Table 9.1).

To construct the database for comparative urinary proteomics universally available, the method of sample collection, storage and protein isolation should be standardized [38] and the recommended one has been presented on the HKUPP Web site. Thus, urine collection programs under the standardized procedures have recently started "Biobank" on a multi-center basis in many countries and also in Japan where the Japanese Society of Nephrology has been leading the program under the subcommittee on a combination of urinary biomarkers and the program of standardized urine collection.

9.2.2.2 Proteomic analysis of urinary exosomes and its problems

The exosomes are small vesicles consisting of plasma membranes with a diameter of 40–100 nm deriving from tubular epithelial cells and partly podocytes and are obtained as low-density vesicles by sequential ultracentrifuge of urine at 17,000g and 200,000g [39, 40]. Knepper's group has been intensively performing the large-scale proteomic analysis for urinary exosomes [41, 42]. In 2004, Pisitkun *et al.* showed that urinary exosomes exist with apical membranes being inside in contrast to the inside-out endosomes and contain membrane proteins such as various transporters [43]. Currently, the databases of human urine exosomes along with protocols of isolation are available on the Web. Gonzarez *et al.* created large-scale phosphorylated proteomics of urinary exosomes by MS, identifying 1132 proteins derived from epithelial cell membranes, suggesting that this approach could be applied to making a diagnosis of some genetic renal diseases [44]. Serious problems encountered in the urine exosome proteomics may be insufficient knowledge about their origins and mechanisms of the release in the urine. These are significant for evaluating the exosome proteome in both physiologic and pathologic conditions.

9.2.3 Blood (Plasma, Serum)

Plasma is one of major targets for proteomic since the analysis is expected to discover biomarkers for various diseases or cancers. Blood or plasma is sometimes utilized for proteomic analysis searching for discovery of possible biomarkers of various renal diseases such as kidney and urinary tract malignancies and also some kidney diseases. The major problem for identification of minor protein components in plasma is the existence of large amounts of several proteins, such as albumin and globulins in plasma that may hinder detection of tiny amounts of proteins in MS analysis. To remove such abundant proteins, several types of columns that will efficiently adsorb the unnecessary ones have been developed. Recently, Salant's group has discovered phsospholipase A2 receptor (PLA2R) as a possible causative antigen responsible for membranous nephropathy, a type of immune complex glomerular disease of unknown mechanism [45]. They made a screening of the proteins obtained from human glomerulus that could bind to the autoantibodies isolated from serum of the patients, and finally identified PLAR2 by MS.

9.2.4 Renal Cells

Proteomic analysis targeting specific cell types of the renal nephron compartments could more accurately elucidate the precise mechanism underlying the disease. In addition to just using primary cultured cells or cell lines *in vitro*, recent advances in cell sorting technique have allowed selective isolation of cells from urine for proteomics approaches. Among studies using culture cells, Rao *et al.* identified calmodulin as a high glucose–dysregulated protein from cytosolic and membrane subproteome of rat and human glomerular mesangial cells in culture [46]. Schordan *et al.* showed that exposure to high glucose altered the expression profiles of some proteins such as cytoskeletal proteins and specific annexins in a podocyte cell line by using MALDI-TOF-MS [47]. Harita *et al.* conducted the proteome analysis to investigate the mechanisms of phosphorylation of nephrin or other slit membrane proteins in the cultured podocytes, clarifying the specific phosphorylation sites by Fyn [48]. Dhihazi *et al.* surveyed the proteome in cultured rabbit thick ascending limb of Henle (TALH) cell lines, showing that cytoskeleton proteins, including vimentin, cytokeratin, and heat shock proteins such as

HSP70, were induced by hyperosmolarity [49]. Recently, Da Silva and Benton reported that in EGFR (epidermal growth factor receptor) transgenic mouse, the intercalated cells expressing vacuolar-type ATPase B1 were successfully collected by FACS, which were subject to subsequent proteomic analysis, suggesting the possibility of using selectively isolated cells for proteomic study [50].

9.3 Applications of Proteomics to Various Renal Diseases

The methodology for proteomic analysis shown above has been applied to clinical and basic investigation attempting to clarify the pathophysiology and discovering useful biomarkers in various renal diseases, including acute kidney injury (AKI), chronic kidney disease (CKD), diabetic nephropathy, primary and secondary glomerulonephritis, and graft rejection as described below.

9.3.1 Acute Kidney Injury (AKI)

Despite recent advances in the understanding of its pathophysiology, the incidence and prognosis of acute renal failure have not been improved in the last decades [51]. Thus, a new entity called acute kidney injury (AKI) has been advocated, aiming at a better survival rate particularly by an earlier diagnosis and intervention. Although a number of biomarkers, including cystatin C, neutrophil gelatinase-associated lipocalin (NGAL), interleukin-18 (IL-18), and kidney injury molecule 1 (KIM 1), are emerging, the efforts for finding novel ones and also a better combination of these markers have been vigorously made taking advantage of proteomics and other new approaches [52, 53].

9.3.1.1 Animal models

Charlwood *et al.* found mitochondrial proteins and apoptosis-associated molecules by analyzing the proteome of renal cortex in rats with AKI induced by gentamycin [54]. Zhao *et al.* investigated urinary exosomes by 2-D DIGE in a rat model of cisplatin-induced AKI, showing that urinary fetuin-A levels significantly increased two days before the increase in serum creatinine concentrations but not in rats with prerenal azotaemia [55]. The usefulness of fetuin-A as an early AKI biomarker has been confirmed in a rat ischemic-reperfusion injury model and also in patients with AKI in the intensive care unit.

9.3.1.2 Clinical studies

Ho *et al.* analyzed urine proteome by SELDI-TOF-MS revealing that a peak of hepsidin-25 was seen only in patients not associated with AKI after cardiac surgery [56]. Bennet *et al.* showed that urine proteome contained an unknown peak in patients developing contrast media-induced AKI, whereas the peak of beta-defensin was not present in patients with AKI, suggesting a reno-protective role of beta–defensin as well as a predictive marker [57]. Metzger *et al.* examined the AKI-specific proteomic profiles in urine by CE-MS and showed that development of AKI could be better predicted by a combination of increased proteins (beta2-microglobulin, albumin, alpha1-microglobulin) and decreased proteins (fibrinogenα and collagens type1α(I) and (III)), than already known biomarkers such as KIM-1, IL-18, cystatin C and NGAL [58]. Such combined approach for serum biomarkers and urine microscopic profiles was also shown to be useful in predicting AKI outcomes [59].

9.3.2 Tubular Disorders

There are a limited number of proteomic studies for tubular disorders and electrolyte imbalance.

As described earlier, urine exosomes are useful for the diagnosis of some genetic renal diseases such as Bartter syndrome, in which the Na,K,2Cl cotransporter was shown to be absent in urinary exosomes in contrast to normal subjects [44]. Cutillas *et al.* used MALDI-TOF-MS and LC-MS revealing that urine protein profiles in the patients with Dent's disease, a hereditary tubular proteinuria caused by a mutation of chloride channel CLC5, were distinct from those of control subjects. High-molecular-weight proteins such as megalin, cubulin, EGF, and uromodulin were absent in the urine of the patients, whereas other proteins such as apoproteins and various carrier proteins were increased [60]. Vilasi *et al.* compared urine proteomes among three hereditary tubular diseases, i.e., Dent's disease, Lowe syndrome (also known as Dent-2 disease) and autosomal dominant Fanconi syndrome of unknown origin, by LC-ESI MS combined with nuclear magnetic resonance (NMR).

They reported that Dent's and Lowe diseases had similar profiles suggesting the mechanistic correlations, whereas the profiles were different in the autosomal dominant Fanconi syndrome [61].

9.3.3 Diabetic Nephropathy

Diabetic nephropathy is currently the leading cause of end-stage renal disease in many countries probably reflecting the increasing population of diabetes for the last several decades. Treatment suppressing or even preventing development of diabetic nephropathy is urgently needed, and various clinical/basic studies have been vigorously attempted for the purpose of detecting the molecules that are pathophysiologically important and/or of predictive values. The biomarkers reported thus far include the decreased levels of type IV collagen and E-cadherin as well as the increased levels of uromodullin and serum proteins such as Ca or vitamin D binding protein, various glycated proteins and some defense-response proteins (see below) [62, 63]. Those markers were identified mainly by urinary proteomic analysis since 2004, when Mischak first reported on the urine proteomes in patients with type 2 diabetes [64]. Proteomic studies using blood samples have been rarely conducted, but Han *et al.* recently identified an increase in glycoproteins such as lumican, vesicant, and retinol-binding protein-4 using multi-lection affinity chromatography and LC-MS/MS [65]. Thus, there are a variety of reports showing different results in diverse diabetic populations, and it will be important to carefully select such markers that are actually relevant and specific among those proteins proposed.

In basic studies, Thongboonkard *et al.* examined proteomes of renal tissues in transgenic OVE26 mice in 2004, showing an increase in elastin and a decrease and elastase IIB, respectively [66]. Schnauzer *et al.* showed the increased levels of pro-alpha-collagen in streptozotocin diabetic rats by LC-MS/MS analysis [67].

There are a number of clinical studies for urinary proteomes in diabetic patients mostly for the purpose of searching for early biomarkers [68, 69]. Rao *et al.* examined urine protcomics of diabetic patients with or without nephropathy after immunodepletion of abundant urinary proteins [70], showing a rise in the levels of urinary defense-response proteins, several glycoproteins and vitamin D-binding proteins according to the development of nephropathy, the latter being similar with the previous report by Thongboonkerd.

Sharma *et al.* identified alpha1-antitrypsin in the urine of type 1 and 2 diabetic patients with overt nephropathy using CE and SELDI-TOF-MS [71]. Later, fragments of collagens are identified as a characteristic peak for diabetic nephropathy using MALDI-TOF-MS and CE-MS [72, 73]. Jiang *et al.* also showed the increased levels of urinary E-cadherin in the urine using DIGE/MALDI-TOF-MS [74]. Rossing *et al.* have intensively studied urinary proteome to detect potential biomarkers in the patients with type 1 and type 2 diabetes and found the characteristic profiles and some specific proteins [75, 76]. They showed that fragments of type 1 collagen were decreased in diabetic patients but not in non-diabetic subjects with ischemic heart disease, suggesting the usefulness of urine proteome as a tool for differential diagnosis.

Recently, longitudinal studies for urine proteomes have also been accumulating aimed at a discovery of diagnostic or predictive biomarkers in diabetic nephropathy. Otu *et al.* followed up Pima Indians with type 2 diabetes without albuminuria for 10 years, identifying unknown peaks only in those developing nephropathy [77]. Andersen and Rossing reported in a substudy of IRMA-2 for type 2 diabetic patients with nephropathy that the diabetes-specific proteome patterns were recovered by treatment with candesartan [78]. Merchant *et al.* performed the follow-up study for urine proteomes of type 1 diabetic patients with microalbuminuria for 10–12 years, showing that declining renal function may be predicted by a combination of three decreased peptides (fragments of type 4 and type 5 collagens and tenascin-X) and three increased peptides (fragments of IP2-kinase, zona occludens 3 and FAT tumor suppressor 2) [79], suggesting the usefulness of combined approach.

9.3.4 IgA Nephropathy and Other Glomerular Diseases

Proteomic analysis has been applied to various glomerular diseases, including IgA nephropathy. Although renal biopsy is a gold standard for a diagnosis and a prediction of clinical courses and therapeutic responsiveness, it is an invasive technique with unignorable risk; thus, new biomarkers in the urine or serum will be warranted.

In IgA nephropathy, a number of studies have been reported, beginning with simply analyzing urinary protein profiles of the patients [80]. Haubitz *et al.* examined urinary proteomes by CE-MS showing that specific patterns of 22 to 28 proteins were seen with a good discrimination from other glomerular disease and healthy subjects [81]. Julian *et al.* also found the characteristic patterns for IgA nephropathy and purpura nephritis, both of which were distinct from those of healthy controls or IgA-related glomerulosclerosis in cirrhosis [82]. Rocchetti *et al.*, by analyzing urine proteome in IgA nephropathy patients, found a set of proteins (kininogen, inter-alpha-trypsin-inhibitor heavy chain 4 and transthyretin), which could discriminate the subjects with a good response to angiotensin converting enzyme inhibitor from unresponders [83]. Kaneshiro *et al.* examined the proteomic profiles of the serum short peptides by using ion exchange beads and MALDI-MS/MS in patients with IgA nephropathy and revealed the characteristic patterns with particular five peptide peaks, which may be related to the pathogenesis of IgA nephropathy [84].

In other primary glomerular diseases, Varghese *et al.* examined urinary proteomes in various glomerular diseases, showing that the urine biomarkers were useful in diagnosing the underlying diseases [85]. Candiano *et al.* analyzed the urine and serum of 23 patients with various idiopathic nephrotic syndrome by MALDI-TOF-MS, revealing protease-induced protein modification or adduct formation in urinary albumin, alpha-1-antitrypsins and apolipoprotein A1, some of which may be derived from serum [86, 87]. Woroniecki *et al.* examined urine proteomes by SELDI-MS in children with idiopathic nephrotic syndrome (NS) showing that the profiles were different between patients with steroid-sensitive and steroid-resistant NS [88]. Musante *et al.* reported that plasma albumin was highly oxidized in focal segmental glomerulosclerosis (FSGS), with cysteine 34 residues substituted by the oxidized form bound to sulfonic group in the relapsing phase, suggesting the involvement of free radicals [89].

In secondary glomerular diseases, Mosley *et al.* identified unknown peaks specific for active lupus nephritis in a cross-sectional analysis for urine proteome [90]. Also, Zhang *et al.* examined urinary proteomic profiles using weak cation exchange chips and SELDI-TOF-MS for 145 samples of 19 lupus nephritis patients, showing that the levels of hepcidin 20 were increased and those of hepcidin 25 decreased before or at renal flare, in addition to the increase in fragments of alpha1-antitrypsin and albumin [91]. As described above, Sethi *et al.* used LCM from FFPE biopsy samples of patients with various membranoproliferative glomerulonephritis, including dense deposit disease (DDD, also known as MPGN type II). They showed that precursors of C5-9 complements, particularly C9, were uniformly detected, but immunoglobulin was absent in DDD, whereas complement-regulating factors such as clusterin, vitronectin, factor H regulator 1 (FHR1) were seen in not only DDD but other immune complex-mediated glomerulonephritis [27]. Haubits *et al.* examined urine proteomes for the patients with anti-neutrophil cytoplasmic antibody (ANCA)-associated vasculitis (AAV) by CE-MS, identifying 113 potential biomarkers that were specific for AAV, 58 of which were sequenced and proved to be hemoglobin fragments. The association with disease activity was confirmed for 18 out of those proteins, suggesting the usefulness of such approach as a diagnostic and monitoring tool [92].

9.3.5 Chronic Kidney Disease and Dialysis Patients

Chronic kidney disease (CKD) has been classified into five stages according to the estimated glomerular filtration rate (eGFR) and is also stratified depending on a degree of albuminuria. The populations with CKD, including those on dialysis, are increasing and are known to have high cardiovascular and renal risks even in the early stage of microalbuminuria. Although several biomarkers other than GFR and albuminuria have been advocated, only a few have become utilized in clinical practice [93]. Thus, proteomic analysis is performed mainly aimed at exploring the better and novel biomarkers with diagnostic, predictive, and pathophysiological relevance. In addition, the analysis of uremic toxins in pre-dialysis and dialysis patients has been conducted by use of proteomics [94].

In animal studies, Xu *et al.* analyzed glomerular proteomes in the 5/6 nephrectomized rats, showing that sclerotic glomeruli could be clearly distinguished from nonsclerotic ones by glomerular proteomic patterns [24]. Later, Porthoff *et al.* examined glomerular proteins from 5/6 nephrectomized rats by LC-ESI-MS, showing an increase in galectin-1 and dimethylarginine dimethylaminohydrolase 1 (DDAH) and a decrease in merprin-alpha in sclerotic glomeruli [95]. Toyohara *et al.* analyzed urine and plasma proteome by CE-MS/HPLC in transgenic rats for organic transporters in renal proximal tubules [96]. Uremic toxins such as guanidinosuccinate (GSA), asymmetric dimethylarginine (ADMA) and trans-aconitate were decreased, in contrast to indoxyl sulfate and citrulline, as compared to control rats, suggesting the role of the organic transporters in preventing accumulation of those anions. Furthermore, the transporters were shown to be upregulated by HMG-CoA reductase inhibitors and so may be a promising candidate for the treatment of uremia.

In clinical studies for pre-dialysis diseases, Zurbig *et al.* analyzed low-molecular-weight urinary proteomes in CKD patients of various ages, showing that high resemblance between aging kidney and CKD; thus, it is suggested that the reduced proteolytic activity may be involved in human renal aging [97]. Good *et al.* have constructed the large-scale database of urine proteome using CE-MS in 3600 subjects with a wide variety of glomerular, vascular, and tubular renal diseases, revealing the characteristic patterns for CKD patients [98].

For dialysis patients, Langlois *et al.* showed using SELDI-TOF-MS that serum peptide proteomes were considerably different in patients on hemodialysis from those with healthy controls [99]. Dihazi *et al.* compared dialysis efficiency between high-flux or low-flux membranes for hemodialysis and peritoneal dialysis (PD) by using 2D-PAGE/SELDI-TOF-MS, showing the differential protein profiles, with toxic middle molecules more efficiently removed by high-flux filters [100]. Molina *et al.* analyzed protein profiles for dialysis effluents using LC-MS/MS, identifying novel 292 proteins that had not been known as serum uremic toxins [101]. Tomosugi *et al.* found using SELDI-TOF-MS the characteristic peaks of hepcidin in the serum of hemodialysis patients by SELDI-TOF-MS analysis and developed the method of quantifying hepcidin 25 [102]. Kemna *et al.* also used SELDI-TOF-MS to measure the concentrations of hepcidin in serum and urine [103]. In continuous ambulatory PD (CAPD) patients, Sritippayawan *et al.* analyzed protein profiles of peritoneal dialyzate using 2D-PAGE and ELISA, showing that the levels of the particular five proteins varied depending on the transport characteristics of peritoneal membranes determined by peritoneal efficiency test (PET) [104]. Interestingly, the levels of kappa light chains were continuously high after episodes of peritonitis.

9.3.6 Renal Allograft Rejection

To prevent graft loss after renal transplantation, it is crucial to make a correct diagnosis of acute and chronic rejection as early as possible, which may allow prompt interventions leading to improved graft survival. Although renal biopsy is a gold standard for the diagnosis, it is associated with a risk of complications; thus, the development of non-invasive approaches is warranted [105]. Urine biomarkers have been investigated since 2004, when Schaub *et al.* reported the comprehensive proteome analysis of the urine in post-transplant patients with acute graft rejections. They found fragments of beta-2 microglobulin in the urine of the affected patients [106, 107], but later it turned out that it was not necessarily specific for graft rejection but rather a universal marker for tubular injury. SELDI-TOF-MS analysis by O'Riordan *et al.* revealed two specific peptide peaks for acute rejection which were subsequently identified as alpha-1-antichymotrypsin and a variant of beta-defensin 1 [108]. Recently, Ling *et al.* analyzed urinary peptidomics using MALDI-TOF-MS with multiple reaction monitoring, suggesting that a panel of proteins that increased (collagens type 1 and 3, MMP-7 and serpin G1) and decreased (uromodullin) may be a predictive biomarker with a high specificity and sensitivity [109]. This was confirmed later by Signdel using LC-MS/MS [110]. Metzger *et al.* identified collagen fragments as an early marker for acute rejections by an examination of urinary proteomics and showed the involvement of MMP-8, which is a molecule reflecting an infiltration of T-cells into the peritubular capillaries in the initial phase of acute rejection [111].

In chronic allograft rejection, SELDI-TOF-MS analysis of urine proteome revealed characteristic peaks, one of which proved to be endorepellin, C-terminal fragments of perlecan [112], and furthermore were shown to be increased prior to an increase in serum creatinine. Quintana showed that chronic active antibody-mediated rejection and pure interstitial fibrosis /tubular atrophy could be well discriminated by urinary proteomic profiles, among which the two specific peptides, uromodullin and kininogen, were particularly useful with high specificity [113, 114].

References

1. Smith, MPW, Banks, RE, Wood, SL, Lewington, JP, and Selby, PJ (2009). Application of proteomic analysis to the study of renal disease. *Nat. Rev. Nephrol.* 5, 701–712.
2. Thongboonkerd, V, and Malasit, P (2005). Renal and urinary proteomics: current applications and challenges. *Proteomics* 5, 1033–1042.

3. O'Riordan, E, Gross, S S, and Goligorsky, MS (2006). Technology insight: renal proteomics—at the crossroads between promise and problems. *Nat. Clin. Pract. Nephrol.* 2, 445–458.
4. Yamamoto, T, Langham, RG, Ronco, P, *et al.* (2008). Toward standard protocols and guidelines for urine proteomics. *Proteomics* 8, 2156–2159.
5. Vlahou, A, Schanstra, J, Frokiaer, J, El Nahas M, *et al.* (2008). Establishment of a European Network for urine and kidney proteomics. *J. Proteomics* 8, 2156–2159.
6. Zhang, Y, Zhang, Y, Adachi, J, Olsen, JV, *et al.* (2007). MAPU: Max-Plank Unified database of organellar, cellular, tissue and body fluid proteomes. *Nucleic. Acids. Res.* 35, D771–D779.
7. Charonis, A, Luider, T, Baumann, M, Schanstra, JP (2011). Is the time ripe for kidney tissue proteomics? *Proteomics Clin. Appl.* 5, 215–221.
8. Yoshida, Y, Miyazaki, K, Kamie, J, Sato, M, *et al.* (2005). Two-dimensional electrophoretic profiling of normal human kidney glomerulus proteome and construction of an extensible markup language (XML)-based database. *Proteomics* 5, 083–1096.
9. Miyamoto, M,*et al.* (2007). In-depth proteomic profiling of the normal human kidney glomerulus using two-dimensional protein prefractionation in combination with liquid chromatography-tandem mass spectrometry. *J. Proteome Res.* 6, 3680–3690.
10. Witzmann, FA, Fultz, CD, Grant, RA, Wright, LS, *et al.* (1998). Differential expression of cytosolic proteins in the rat cortex and medulla: preliminary proteomics. *Electrophoresis* 19, 2491–2497.
11. Sitek, B, Potthoff, S, Schulenborg, T, Stegbauer, J, *et al.* (2006). Novel approaches to analyse proteins from smallest scale murine and human samples using DIGE saturation labeling. *Proteomics* 6, 4337–4345.
12. Zhao, Y, Denner, L, Haidacher, SJ, LeJeune, WS, and Tilton, RG (2008). Comprehensive analysis of the mouse renal cortex using two-dimensional HPLC-tandem mass spectrometry. *Proteomics Sci.* 6, 1–15.
13. Zhang, Y, Yoshida, Y, Nameta, M, Xu, B, *et al.* (2010). Glomerular proteins related to slit diaphragm and matrix adhesion in the foot process are highly tyrosine phosphorylated in the normal rat kidney. *Nephrol. Dial. Transplant.* 25, 1785–1795.
14. Chou, CL, *et al.* (2004). Non-muscle myosin II and myosin light chain kinase are downstream targets for vasopressin signalling in the renal collecting duct. *J. Biol. Chem.* 279, 49026–49035.

15. Hoorn, EJ, Hoffert, JD, and Knepper, MA (2005). Combined proteomics and pathways analysis of collecting duct reveals a protein regulatory network activated in vasopressin escape. *J. Am. Soc. Nephrol.* 16, 2852–2863.

16. Pisitkun, T, Bieniek, J, Tchapyjnikov, D, Wang, G, *et al.* (2006). High-throughout identification of IMCD proteins using LC-MS/MS. *Physiol. Genomics* 25, 263–276.

17. Hoffert, JD, Pisitkun, T, Wang, G, Shen, R, and Knepper, MA (2006). Quantitative phosphoproteomics of vasopressin-sensitive renal cells: regulation of aquaporin-2 phosphorylation at two sites. *Proc. Natl. Acad. Sci. U S A* 103, 7159–7164.

18. Hoffert, JD, Wang, G, Pisitkun, T, Shen, RF, and Knepper, MA (2007). An automated platform for analysis of phosphoteomic datasets: application to kidney collecting duct phosphoproteins. *J. Proteome Res.* 6, 3501–3508.

19. Nielsen, J, *et al.* (2008). Proteomic analysis of lithium-induced nephrogenic diabetes insipidus: mechanisms for aquaporin 2 down-regulation and cellular proliferation. *Proc. Natl. Acad. Sci. USA* 105, 3634–3639.

20. Rivard, CJ, *et al.* (2007). Expression of the calcium-binding protein S100A4 is markedly up-regulated by osmotic stress and is involved in the renal osmoadaptive response. *J. Biol. Chem.* 282, 6644–6652.

21. Curthoys, NP, Taylor, L, Hoffert, JD, Knepper, MA (2007) Proteomic analysis of the adaptive response of rat renal proximal tubules to metabolic acidosis. *Am. J. Physiol.* 292, F140–F147.

22. Xu, BJ (2010). Combinging laser capture mircodissection and proteomics: methodologies and clinical applications. *Proteomics Clin. Appl.* 4, 116–123.

23. Xu, BJ, Syr, Y, Liang, X, Ma, LJ, *et al.* (2005). Proteomic patterns and prediction of glomerulosclerosis and its mechanisms. *J. Am. Soc. Nephrol.* 16, 2967–2975.

24. Wiśniewski, JR, Ostasiewicz, P, Mann, M. (2011). High recovery FASP applied to the proteomic analysis of microdissected formalin fixed paraffin embedded cancer tissues retrieves known colon cancer markers. *J. Proteome Res.* 10, 3040–3049.

25. Xu, B, Shimada, S, Zhang, Y, Magdeldin, S, Fujinaka, H, Yoshida, Y, Yaoita, Y, Yamamoto, T (2011). Comparison of on site proteomics study on laser microdissected FFPE and frozen human glomerulus tissues. 10th Annual World Congress of Human Proteome Organization (HUPO), OC055, in September 2011, Geneva, Switzerland.

26. Sethi, S, Gamez, JD, Vrana, JA, Theis, JD, *et al.* (2009). Glomeruli of dense deposit disease contain components of the alternative and terminal complement pathway. *Kidney Int.* 75, 952–96.

27. Sedor, JR (2009). Tissue proteomics: a new investigative tool. *Kidney Int.* 75, 876–879.

28. Waanders, LF, Chwalek, K, Monetti, M, Kumar, C, Lammert, E, and Mann, M (2009). Quantitative proteomic analysis of single pancreatic islets. *Proc. Natl. Acad. Sci. U S A* 10, 18902–18907.

29. DeSouza, LV, Krakovska, O, Darfler, MM, Krizman, DB, *et al.* (2010). mTRAQ-based quantification of potential endometrial carninoma biomarkers from archibed formali. ne-fixed paraffin-embedded tissues. *Proteomics* 10, 3108–3116.

30. Thongboonkerd, V. (2007). Practical points in urinary proteomics. *J. Proteome Res.* 6, 3881–3890.

31. Thongboonkerd, V, and Saetum, P, (2007). Bacterial overgrowth affects urinary proteome analysis: recommendation for centrifugation, temperature, duration, and the use of preservatives during sample collection. *J. Proteome Res.* 6, 4173–4181.

32. Zurbig, P, Dihazi, H, Metzger, J, Thongboonkerd, Vlahou, A (2011). Urine proteomics in kidney and urogenital diseases: Moving towards clinical applications. *Proteomics Clin. Appl.* 5, 256–268.

33. Mischak, H, Kolch, W, Avivalotis, M, Bouyssie, D, *et al.* (2010). Comprehensive human urine standards for comparability and standarization in clinical proteome analysis. *Proteomics Clin. Appl.* 4, 464–478.

34. Pieper, R, *et al.* (2004). Characterization of the human urinary proteome: a method for high-resolution display of urinary proteins on two dimensional electrophoresis gels with a yield of nearly 1,400 distinct protein spots. *Proteomics* 4, 1159–1174.

35. Chen, Y, *et al.* (2007). Large-scale protein identification of human urine proteome by multi-dimensional LC and MS/MS. *Proteomics Clin. Appl.* 1, 577–587.

36. Mataija-Botelho, D, *et al.* (2009). A qualitative proteome investigation of the sediment portion of human urine: implications in the biomarker discovery process. *Proteomics Clin. Appl.* 3, 95–105.

37. Adachi, J, Kumar, C, Zhang, Y, Olsen, JV, and Mann, M (2006). The human urinary proteome contains more than 1500 proteins, including a large proportion of membrane proteins. *Genome Biol.* 7, R80.

38. Fiedler, GM, *et al.* (2007). Standardized peptidome profiling of human urine by magnetic bead separation and matrix-assisted laser desorption/ionization time-of-flight mass spectrometry. *Clin. Chem.* 53, 421–428.

39. Sachs, AN, Pisitkun, T, Hoffert, JD, Yu, MJ, *et al.* (2008). LC-MS/MS analysis of differential centrifugation fractions from native inner medullary collecting duct of rat. *Am. J. Physiol. Renal Physiol.* 295, F1799–F1806.

40. Yu, MJ, Pisitkun, T, Wang, G, Aranda, JF, *et al.* (2008). Large scale quantitative LC-MS/MS analysis of detergent-resistant membrane proteins from rat renal collecting duct. *Am. J. Physiol. Cell Physiol.* 295, C661–C678.

41. van Balkom, BWM, Pisitkun, T, Verhaar, M, Knepper, MA (2011). Exosomes and the kidney: prospects for diagnosis and therapy of renal diseases. *Kidney Int.* 80, 1138–1145.

42. Gonzalez, P, Pisitkun,T, Knepper, MS (2008). Urinary exosomes: is there a future? *Nephrol. Dial. Transplant.* 23, 1799–1801.

43. Pisitkun, T, Shen, R, and Knepper, MA (2004). Identification and proteomic profiling of exosomes in human urine. *Proc. Natl. Acad. Sci. U S A* 101, 13368–13373.

44. Gonzales, PA, *et al.* (2009). Large-scale proteomics and phosphoproteomics of urinary exosomes. *J. Am. Soc. Nephrol.* 20, 363–379.

45. Beck, Jr, LH, Bonegio, RGB, Lambeau, G, Beck, DM, *et al.* (2009). M-type phospholipase A2 receptor as target antigen in idiopathic membranous nephropathy. *New Engl. J. Med.* 361, 11–21.

46. Ramachandra, Rao, SP, Wessell, R, Shaw, A, and Sharma, K (2007). Profiling of human mesangial cell subproteomes reveals a role for calmodulin in glucose uptake. *Am. J. Physiol. Renal Physiol.* 292, F1182–F1189.

47. Schordan, S, Schordan, E, Endlich, Lindenmeyer, MT, *et al.* (2009). Alterations of podocyte proteome in response to high glucose concentrations. *Proteomics* 9, 4519–4528.

48. Harita, Y, Kurihara, H, Kosako, H, Tezuka, T, *et al.* (2009). Phosphorylation of nephrin triggers Ca^{2+} signaling by recruitment and activation of phospholipase C-gamma 1. *J. Biol. Chem.* 284, 8951–8962.

49. Dihazi, H., Asif, AR, Agarwal, NK, Doncheva, Y, and Muller, GA (2005). Proteomic analysis of cellular response to osmotic stress in thick

ascending limb of Henle's loop (TALH) cells. *Mol. Cell. Proteomics* 4, 1445–1458.

50. Da Silva, N, Pisitkun, T, Belleannée, C, Miller, LR, Nelson, R, Knepper, MA, Brown, D, Breton, S (2010). Proteomic analysis of V-ATPase-rich cells harvested from the kidney and epididymis by fluorescence-activated cell sorting. *Am. J Physiol. Cell Physiol.* 298, C1326–C1342.
51. Murugan, R, and Kellum, JA (2011). Acute kidney injury: what's the prognosis. *Nat. Rev. Nephrol.* 7, 209–217.
52. Siew, ED, Ware, L, and Ikizler, TP (2011) Biological markers of acute kidney injury. *J. Am. Soc. Nephrol.* 22, 810–820.
53. Coca, SG, Yalavarthy, R, Concato, J, and Parikh, CR (2008). Biomarkers for the diagnosis and risk stratification of acute kidney injury: a systematic review. *Kidney Int.* 73, 1008–1016.
54. Charlwood, J, Skehel, JM, King, N, Camilleri, P, *et al.* (2002). Proteomic analysis of rat kidney cortex following treatment with gentamicin. *J. Proteome Res.* 1, 73–82.
55. Zhou, H, *et al.* (2006). Exosomal Fetuin-A identified by proteomics: a novel urinary biomarker for detecting acute kidney injury. *Kidney Int.* 70, 1847–1857.
56. Ho, J, Lucy, M, Kronkhin, O, Hayglass, K, *et al.* (2009). Mass spectrometry-based proteomic analysis of urine in acute kidney injury following cardiopulmonary bypass: a nested case-control study. *Am. J. Kidney Dis.* 53, 584–595.
57. Bennett, MR, *et al.* (2008). Using proteomics to identify preprocedural risk factors for contrast induced nephropathy. *Proteomics Clin. Appl.* 2, 1058–1064.
58. Metzger, J, Kirsch, T, Schiffer, E, Ulger, P, Mentes, E, *et al.* (2010). Urinary excretion of twenty peptides forms an early and accurate diagnostic pattern of acute kidney injury. *Kidney. Int.* 78, 1252–1262.
59. Hall, IE, Coca, SG, Perazella, MA, Eko, UU, *et al.* (2011). Risk of Poor Outcomes with Novel and Traditional Biomarkers at Clinical AKI Diagnosis. *Clin. J. Am. Soc. Nephrol.* 6, 2740–2749.
60. Cutillas, PR, *et al.* (2004). The urinary proteome in Fanconi syndrome implies specificity in the reabsorption of proteins by renal proximal tubule cells. *Am. J. Physiol. Renal Physiol.* 287, F353–F364.
61. Vilasi, A, *et al.* (2007). Combined proteomic and metabonomic studies in three genetic forms of the renal Fanconi syndrome. *Am. J. Physiol. Renal Physiol.* 293, F456–F467.

62. Mischak, H, Rossing, P (2010). Proteomic biomarkers in diabetic nephropathy-reality or future promise? *Nephrol. Dial. Transplant.*25, 2843–2845.

63. Ameur, RB, Molina, L, Bolvin, C, Kifagi, C, *et al.* (2010). Proteomic approaches for discovering biomarkers of diabetic nephropathy. *Nephrol. Dial. Transplant.* 25, 2866–2875.

64. Mischak, H, *et al.* (2004). Proteomic analysis for the assessment of diabetic renal damage in humans. *Clin. Sci.* 107, 485–495.

65. Ahn, JM, Kim, BG, Yu, MH, *et al.* (2010). Identification of diabetic nephropathy-selective proteins in human plasma by multi-lectin affinity chromatography and LC-MS/MS. *Proteomics Clin. Appl.* 4, 644–653.

66. Thongboonkerd, V, *et al.* (2004). Alterations in the renal elastin-elastase system in type I diabetic nephropathy identified by proteomic analysis. *J. Am. Soc. Nephrol.* 15, 650–662.

67. Schlatzer, DM, *et al.* (2009). Urinary protein profiles in a rat model for diabetic complications. *Mol. Cell. Proteomics* 8, 2145–2158.

68. Good, DM, Zurbig, P, Argiles, A, Bauer, HW, *et al.* (2010). Naturally occurring human urinary peptides for use in diagnosis of chronic kidney disease. *Mol. Cell. Proteomics* 9, 2424–2437.

69. Bellei, E, *et al.* (2008). Proteomic analysis of early urinary biomarkers of renal changes in type 2 diabetic patients. *Proteomics Clin. Appl.* 2, 478–491.

70. Rao, PV, Lu, X, Standley, M, Pattee, P, *et al.* (2007). Proteomic identification of urinary biomarkers of diabetic nephropathy. *Diabetes Care* 30, 629–637.

71. Sharma, K, *et al.* (2005). Two-dimensional fluorescence difference gel electrophoresis analysis of the urine proteome in human diabetic nephropathy. *Proteomics* 5, 2648–2655.

72. Lapolla, A, *et al.* (2009). Low molecular weight proteins in urines from healthy subjects as well as diabetic, nephropathic and diabetic-nephropathic patients: a MALDI study. *J. Mass Spectrom.* 44, 419–425.

73. Alkahalaf, A, Zurbig, P, Bakker, SJ, Bilo, HJ, *et al.* (2010). Multicentric validation of proteomic biomarkers in urine specific for diabetic nephropathy. *PLoS ONE* 5, e13421.

74. Jiang, H, *et al* (2009). Identification of urinary soluble E-cadherin as a novel biomarker for diabetic nephropathy. *Diab. Metab. Res. Rev.* 25, 232–241.

75. Rossing, K, Mischak, H, Parving, HH, Christensen, PK, *et al.* (2005). Impact of diabetic nephropathy and angiotensin II receptor blockade on urinary polypeptide patterns. *Kidney Int.* 68, 193–205.
76. Rossing, K, Mischak, H, Dakna, M, Zurbig, P, *et al.* (2008). Urinary proteomics in diabetes and CKD. *J. Am. Soc. Nephrol.* 19, 1283–1290.
77. out, HJ, Can, H, Spentzos, D, Nelson, RG, *et al.* (2007). Prediction of diabetic nephropathy using urine proteomic profiling 10 years prior to development of nephropathy. *Diabetes Care* 30, 683–643.
78. Andersen, S, Mischak, H, Zurbig, P, Parving, HH, Rossing, P (2010). Urinary proteome may predict renal function decline in type 1 diabetes and microalbuminuria. *J. BMC Nephrol.* 11, 29.
79. Merchant, ML, *et al.* (2009). Urinary peptidome may predict renal function decline in type 1 diabetes and microalbuminuria. *J. Am. Soc. Nephrol.* 20, 2065–2074.
80. Park, M, *et al.* (2006). Establishment of a 2-D human urinary proteomic map in IgA nephropathy. *Proteomics* 6, 1066–1076.
81. Haubitz, M, *et al.* (2005). Urine protein patterns can serve as diagnostic tools in patients with IgA nephropathy. *Kidney Int.* 67, 2313–2320.
82. Julian, BA, Wittke, S, Novak, J, Good, DM, *et al.* (2007). Electrophoretic methods for analysis of urinary polypeptides in IgA-associated renal diseases. *Electrophoresis* 28, 4469–4483.
83. Rocchetti, MT, Centra, M, Papale, M, Bortone, G, Palermo, C, *et al.* (2008). Urine protein profile of IgA nephropathy patients may predict the response to ACE-inhibitor therapy. *Proteomics* 8, 206–216.
84. Kaneshiro, N, Xiang, Y, Nagai, K, Kurokawa, MS, *et al.* (2009) Comprehensive analysis of short peptides in sera from patients with IgA nephropathy. *Rapid Commun. Mass. Spectrom.* 23, 3720–3728.
85. Varghese, SA, Powell, TB, Budisavjevic, MN, Oates, JC, *et al.*(2007). Urine biomarkers predict the cause of glomerular disease. *J. Am. Soc. Nephrol.* 18, 913–922.
86. Candiano, G, *et al.* (2006). Repetitive fragmentation products of albumin and α1-antitrypsin in glomerular diseases associated with nephrotic syndrome. *J. Am. Soc. Nephrol.* 17, 3139–3148.
87. Candiano, G, *et al.* (2008). Proteins and protein fragments in nephrotic syndrome: clusters, specificity and mechanisms. *Proteomics Clin. Appl.* 2,956–963.
88. Woroniecki, RP, *et al.* (2006). Urinary proteome of steroid sensitive and steroid resistant idiopathic nephrotic syndrome of childhood. *Am. J. Nephrol.* 26, 258–267.

89. Musante, L, *et al.* (2007). Active focal segmental glomerulosclerosis is associated with massive oxidation of plasma albumin. *J. Am. Soc. Nephrol.* 18, 799–810.

90. Mosley, K, *et al.* (2006). Urinary proteomic profiles distinguish between active and inactive lupus nephritis. *Rheumatology* 45, 1497–1504.

91. Zhang, X, *et al.* (2008). Biomarkers of lupus nephritis determined by serial urine proteomics. *Kidney Int.* 74, 799–807.

92. Haubitz, M, Good, DM, Woywodt, A, Haller, H, *et al.* (2009). Identification and validation of urinary biomarkers for differential diagnosis and evaluation of therapeutic intervention in ANCA associated vasculitis. *Mol. Cell. Proteomics* 8, 2296–2307.

93. Fassett, R, Venuthuruoalli, SK, Gobe, GC, Coombes, JS, Cooper, MA, Hoy, WE (2011). Biomarkers in chronic kidney disease: a review. *Kidney Int.* 80, 806–821.

94. Spasovski, G, Ortiz, A, Vanholder, R, El Nahas, M (2011). Proteomics in chronic kidney disease: The issues clinical nephrologists need an answer for. *Proteomics Clin. Appl.* 5, 233–240.

95. Potthoff, SA, *et al.* (2008). The glomerular proteome in a model of chronic kidney disease. *Proteomics Clin. Appl.* 2, 1127–1139.

96. Toyohara, T, Suzuki, T, Morimoto, R, Akiyama, Y, *et al.* (2009). SLCO4C1 transporter eliminates uremic toxins and attenuates hypertension and renal inflammation. *J. Am. Soc. Nephrol.* 20, 2546–2555.

97. Zurbig, P, Decramer, S, Dakna, M, Jantos, J, *et al.* (2009). The human urinary proteome reveals high similarity between kidney aging and chronic kidney disease. *Proteomics* 9, 2108–2117.

98. Good, DM, Zürbig, P, Argilés, A, Bauer, HW, *et al.* (2010). Naturally occurring human urinary peptides for use in diagnosis of chronic kidney disease. *Mol. Cell. Proteomics* 9, 2424–2437.

99. Langlois, RG, *et al.* (2004). Serum protein profile alterations in hemodialysis patients. *Am. J. Nephrol.* 24, 268–274.

100. Dihazi, H, Muller, CA, Mattes, H, and Muller, GA (2008). Proteomic analysis to improve adequacy of hemo- and peritoneal dialysis: Removal of small and high molecular weight proteins with high- and low-flux filters or a peritoneal membrane. *Proteomics Clin. Appl.* 2, 1167–1182.

101. Molina, H, *et al.* (2005). A proteomic analysis of human hemodialysis fluid. *Mol. Cell. Proteomics* 4, 637–650.

102. Tomosugi, N, *et al.* (2006). Detection of serum hepcidin in renal failure and inflammation by using ProteinChip System. *Blood* 108, 1381–1387.

103. Kemna, EH, Tjalsma, H, Podust, VN, and Swinkels, DW (2007). Mass spectrometry-based hepcidin measurements in serum and urine: analytical aspects and clinical implications. *Clin. Chem.* 53, 620–628.

104. Sritippayawan, S, *et al.* (2007). Proteomic analysis of peritoneal dialysate fluid in patients with different types of peritoneal membranes. *J. Proteome Res.* 6, 4356–4362.

105. Sigdel, TK, Lee, S, Sarwal, MM (2011). Profiling the proteome renal transplantation. *Proteomics Clin. Appl.* 5, 269–280.

106. Schaub, S, *et al.* (2004) Proteomic-based detection of urine proteins associated with acute renal allograft rejection. *J. Am. Soc. Nephrol.* 15, 219–227.

107. Schaub, S, *et al.* (2005). Proteomic-based identification of cleaved urinary β2-microglobulin as a potential marker for acute tubular injury in renal allografts. *Am. J. Transplant.* 5, 729–738.

108. O'Riordan, E, *et al.* (2007) Characterisation of urinary peptide biomarkers of acute rejection in renal allografts. *Am. J. Transplant.* **7**, 930–940.

109. Sigdel, T, Kaushal, A, Gritsenko, M, Norbeck, AD, *et al.* (2011). Shotgun proteomics identifies proteins specific for acute renal transplant rejection. *Proteomics Clin. Appl.* 4, 32–47.

110. Ling, XB, Sigdel, TK, Lau, K, Ying, L, *et al.* (2010). Integrative urinary peptidomics in renal transplantation identifies biomarkers for acute rejection. *J. Am. Soc. Nephrol.* 21, 646–653.

111. Metzger, J, Chatzikyrkou, C, Broecker, V, Schiffer, E, *et al.* (2011). Diagnosis of subclinical and clinical T-cell mediated rejection in renal transplant patients by urinary proteomic analysis. *Proteomics Clin. Appl.* 2011.

112. O'Riordan, E, *et al.* (2008). Urinary proteomic analysis of chronic allograft nephropathy. *Proteomics Clin. Appl.* 2, 1025–1035.

113. Quintana, LF, *et al.* (2009). Urine proteomics to detect biomarkers for chronic allograft dysfunction. *J. Am. Soc. Nephrol.* 20, 428–435.

114. Quintana, LF, *et al.* (2009). Application of label-free quantitative peptidomics for the identification of urinary biomarkers of kidney chronic allograft dysfunction. *Mol. Cell Proteomics* 8, 1658–1673.

Chapter 10

Aging-Related Proteome

Tosifusa Toda
Research Team for Mechanism of Aging,
Tokyo Metropolitan Institute of Gerontology,
35-2 Sakaecho, Itabashi, Tokyo 173-0015, Japan
ttoda@tmig.or.jp

Aging is a biological process that occurs in every living organism at the cellular and molecular level. However, the molecular mechanisms of aging are poorly understood yet. Aging-related protein alterations are implicated in the functional deterioration of cells, tissues, and organs in aged individuals. Proteomic investigations into the aging-related protein alterations have been thought to be an effective approach to the molecular mechanisms of aging.

Replicative aging has been studied as a model of cellular aging in which normal mitotic cells may divide a limited number of times, eventually undergoing a growth arrest termed cellular senescence [1–3]. Protein alterations occurring in the replicative aging were first profiled by 2-D gel-based proteome analysis in the normal human diploid fibroblast line TIG-3 [4]. The replicative aging-related variations of protein expression could be categorized into five patterns as a result of the comparative image analysis of protein spots

Quantitative Proteome Analysis: Methods and Applications
Edited by Kazuhiro Imai and Sam F. Y. Li

ISBN 978-981-4316-51-4 (Hardcover), 978-981-4316-98-9 (eBook)
www.panstanford.com

on 2-D gel maps. Among them the most interesting variation was the transitional increase at around 60 PDL (population doubling levels), after that the doubling time of the cell population was extended in a logarithmic scale. The spots of stathmin, Hsp60, Hsp70, TCTP, and SOD1 showed such transitional increase suggesting the involvement in the molecular process of replicative aging. The database have been constructed on the aging-related protein alterations, and opened for free access at the URL http://www.proteome.jp/2D/Fibro/humfb_menu.html.

In the other aspect, terminally differentiated cells in the post-mitotic state show a distinct phenotype of functional cell aging. Damage accumulated in the cells under oxidative stress is implicated in functional cell aging and in geriatric disorders such as vascular and neurodegenerative diseases. The comprehensive analysis was performed on the aging-related protein alterations in mouse brain tissues by Tsugita *et al.* [5] first, in which five regions at five ages from the 10th week to the 24th month were compared by 2-D gel electrophoresis. Over 1,000 protein spots were detected by silver staining, and 17 protein spots were found varying in quantity in the course of aging. Among them, the age-dependent decrease was found in histone H3 spot of p*I* 5.5, suggesting the involvement of the hyper-acetylated form of histone H3 in the mechanism of hippocampal aging.

10.1 Introduction

Biological aging is an inevitable process in which functional deterioration gradually proceeds in almost all tissues and organs throughout the individual life span, especially after maturation. Aging-related alteration appears in both mitotic and post-mitotic cells.

Replicative aging is a typical phenotype of cellular aging in which normal mitotic cells count the number of times they have divided, eventually undergoing a growth arrest termed cellular senescence. The reduced mitotic potential of replicative cells in reproductive tissues may cause age-dependent impairment of tissue regeneration.

The replicative aging of normal human diploid fibroblasts was first found by Hayflick and Moorhead [1, 2] and recommended as a model of cellular senescence. The shortening of telomere, found in the process of replicative aging by Harley *et al.* [3], is the major event counting the number of replication, however, not all the phenotypes of senescent cells, such as reduction in cell migration [6], accumulation of auto-fluorescent proteins and enlargement of cell size, could be ascribed to the shortening of telomere.

The human diploid fibroblast strain TIG-3 [7] established in Tokyo Metropolitan Institute of Gerontology has also a finite life span that is defined by the limited number of times the cell can divide, even under an optimal culture condition, unless otherwise immortalized. The replicative aging of TIG-3 was thought to be useful model for comprehensive profiling of aging-related proteome alterations.

2-D PAGE [8, 9] has been widely used in various fields of research for detecting protein alterations in biological processes such as cell differentiation, immortalization and aging. Celis and co-workers [10] established their original 2-D gel protein database of MRC-5 human embryonal lung fibroblast, which was applied to the quantitative profiling of polypeptides of which relative abundance differed between quiescent, proliferating, and SV40-transformed cells [10]. However, no database describing age-related protein alterations have been established yet. Thus, we decided to construct our original 2-D gel-based proteome database upon replicative aging of TIG-3 [4]. The clickable map of protein spots was prepared from silver-stained 2-D gel images. The data of relative abundance of proteins were linked to the image map.

10.2 Two-Dimensional Gel Electrophoresis (2-DE)-Based Quantitative Proteome Analysis of Replicative Aging

Two-dimensional (2-D) gel electrophoresis-based quantitative proteome analysis is appropriate for screening of protein alterations in both expression level and posttranslational modification appearing in replicative aging of mitotic cells, and functional aging of post-mitotic cells.

In our studies on replicative aging-related proteome, proteins extracted from normal human diploid fibroblast TIG-3 cells at various population doubling levels (PDL), were separated by isoelectric focusing on a strip of immobilized pH gradient (IPG-IEF) in the first dimensional direction, then dissolved by sodium dodecyl sulfate electrophoresis on a polyacrylamide gel slab (SDS-PAGE) in the second-dimensional direction. Protein spots were stained with silver and quantitated by image processing. Corresponding protein spots were matched across all 2-D gel images, and the profiles in quantitative alteration of each protein spots in the course of cellular aging were categorized into five patterns, (i) simple increase, (ii) simple decrease, (iii) decrease after transient increase, (iv) increase after transient decrease, and (v) irregular or no significant variation. The results of the aging related proteome analysis were stored in our original database, TMIG-2DPAGE, in which the information about the relative abundance of each spots determined by the quantitative 2-D gel image analysis was managed.

An alternative "gel-free" method based on LC shotgun MS has been also developed for quantitative proteome analysis [11, 12]. However, major parts of the information about the real protein structure including multiple post-translational modifications have been lost in the process of tryptic digestion. In our preliminary studies, many spots of different p*I* and molecular mass were assigned to a common gene product by Mascot search, suggesting they are processed by multiple post-translational modifications. This was the reason why we decided to use the 2-D gel-based method for analyzing age-related proteome.

10.2.1 Preparation of Cell Proteins for Analyzing Replicative Aging

TIG-3 human fetal lung diploid fibroblasts line established in our institute [7] was employed in the proteomic investigation on replicative aging. The cell line had been confirmed free from mycoplasma contamination. The karyotype of the cell line was analyzed by Matsuo *et al.* [7]. The cells were serially subcultured in Eagle's basal medium (BME, GIBCO) supplemented with 10% fetal bovine serum (FBS) by a standard method of 1:4 splitting throughout the *in vitro* life span. The cell line showed the replicative senescence around 76–80 PDL in the culture condition.

Cells at various PDLs were grown to be 70–80% confluent. After rinsing with phosphate-buffered saline, cells were harvested by scraping with a plastic scraper without using trypsin. The cell suspension was transferred to a microfuge tube, and supplemented with urea (8 M), 2-ME (3%), SDS (0.02%) and Triton X-100 (0.1%

in final concentration). Cells were disrupted by ultrasonication, and the supernatant was removed by centrifugation.

10.2.2 Two-Dimensional Gel Electrophoresis

In our studies on replicative aging, 2-D gel electrophoresis was performed mainly according to the ISO-DALT method with slight modifications reported in our previous paper [4]. The original method of high-resolution two-dimensional gel electrophoresis (2-DE), established by O'Farrell [8], is a powerful tool for resolving hundreds of proteins in a crude sample as isolated spots. The spots can be quantified by image analysis and identified by mass spectrometry. In the first dimension, proteins were resolved by isoelectric focusing (IEF) in accordance with their net charge. In the second dimension, the proteins were further separates by SDS-PAGE depending on their molecular mass.

An improved method of 2-DE, ISO-DALT, established by Görg [9] was performed in our studies on replicative aging-related proteome, in which isoelectric focusing was carried out on a gel strip of immobilized pH gradient in the first dimension (IPG-IEF). IPG-IEF has a critical advantage over the conventional method of isoelectric focusing in which pH gradient was generated using carrier ampholytes. The un-immobilized pH gradient is not stable and drifts toward the cathodic direction in a long term focusing.

After the IPG-IEF gel strip was equilibrated with SDS-containing buffer, and fixed on the top of a slab gel for SDS-PAGE in the second dimension. In the second-dimensional SDS-PAGE, proteins are further separated according to their molecular mass.

10.2.3 Silver Staining and CBB Staining of 2-D Gel

2-D gel-based quantitative proteome research relies on methods for in-gel protein detection. After 2-DE, an appropriate method in visualizing protein spots on a 2-D gel should be carried out for assuring the quantitative analysis. Staining with Coomassie Brilliant Blue (CBB) [13] has been used as the most reliable method for quantitative visualization of protein spots on a 2-D gel. CBB is thought to bind to proteins through electrostatic interactions between the sulfonic groups of the dye and the basic side groups of amino acids [13, 14]. However, the sensitivity of CBB-stain is not high enough for detecting low abundant protein spots on an analytical 2-D gel in which a limited amount of protein sample could be subjected to the analysis. Silver staining [15] has been developed as an alternative method for visualizing proteins in polyacrylamide gel with the significant advantage in sensitivity over CBB staining. The original methods of silver staining, in which glutaraldehyde was used as cross-linking reagent, could not be applicable to gels for mass-spectrometric protein identification. Mass spectrometry (MS)-compatible silver staining [16] has been developed, in which glutaraldehyde treatment was omitted. However, the MS-compatible silver staining compromises sensitivity in detection. Thus, in our research on replicative aging-related proteome, we decided to use the modified procedure of silver staining [17] with slight modifications for achieving the enhanced uniform sensitivity and acceptable linearity in the dynamic range a standard method of silver staining in which glutaraldehyde-based sensitizes in the fixing and sensitization step was performed to achieve an adequate sensitivity and dynamic range for quantitative proteomics.

The 2-D gel, on which higher amount of protein sample was separated for protein identification, was stained with Coomassie Brilliant Blue (CBB).

10.2.4 Quantitative Analysis of Protein Spots by Image Processing

Image analysis of two-dimensional gels is a crucial step in a proteomic workflow and has a direct impact on obtained qualitative and quantitative data. Various types of computer software for achieving the quantitative analysis of protein spots were primarily developed on DEC minicomputers and SUN workstations for institutional/academic use [18–20]. After maturation of the development, some of the software, including PDQuest, used in our research, was released commercially from PDI.

The specific imaging system should be for digitizing the 2-D gel images, visualized by staining with silver, fluorescent dyes or CBB. The silver-stained gel images obtained in our research were acquired by using a flat-bed scanning densitometer and digitized on the PDQuest software. Before the software automatically detects the protein spots on a 2-D gel, the noise in the raw image data was reduced by smoothing; the gel background was subtracted by the "rolling ball" method. PDQuest software models protein spots mathematically as 3-D Gaussian distributions and uses the models to determine absorption maxima. The hit rate of automatic spot detection is highly dependent on the quality of the 2-D gels.

The next step in 2-D gel evaluation is the identification of proteins that are present in all gels of a series of a comparative analysis. Before the software can detect and document matching of different spots, a number of landmarks, or identical spots in the gel series, must be manually identified. The results of the automatic gel comparison require verification, as does automatic spot detection. The quantitative analysis of the aging-related protein variations among gels are performed using the data of relative abundance that was normalized in comparison to the total amount of protein on each gel.

10.2.5 Protein Identification

For protein identification, 2-D gel electrophoresis was performed in a preparative scale using a higher amount of protein sample prepared from pooled TIG-3 cells at various passages of subculture. Protein spots were visualized by CBB staining and matched to silver-stained spots on analytical gel.

The spots on the CBB-stained gel for protein identification were matched to corresponding spots on the silver-stained analytical gel. CBB-stained protein spots were excised and subjected to in-gel tryptic digestion. Alternatively, CBB-stained spots were re-solubilized and transferred to PVDF membrane for on-membrane Lys-C digestion. The monoisotopic masses of peptides, digested with trypsin or Lys-C, were analyzed by using MALDI-TOF MS and ESI-Q-TOF MS systems for protein identification by MASCOT search.

10.2.6 Expression Profiling of Protein in Normal Human Diploid Fibroblasts, TIG-3, with Replicative Aging

Cellular proteins extracted from TIG-3 at 15, 40, 65, 74, 76 and 78 PDL were separated by 2-DE and 468 spots were detected by silver staining. The relative abundance of matched protein spots were quantitatively compared among gels, and the replicative aging-related protein variations were categorized into five patterns: (i) simple increase, (ii) simple decrease, (iii) decrease after transient increase, (iv) increase after transient decrease, and (v) irregular or no significant variation.

The database of 2-D gel based quantitative proteome analysis on the replicative aging-related protein variation was constructed in a Web server, on which categories of the aging-related protein variations were indicated in pseudocolor on the clickable map as shown in Fig. 10.1.

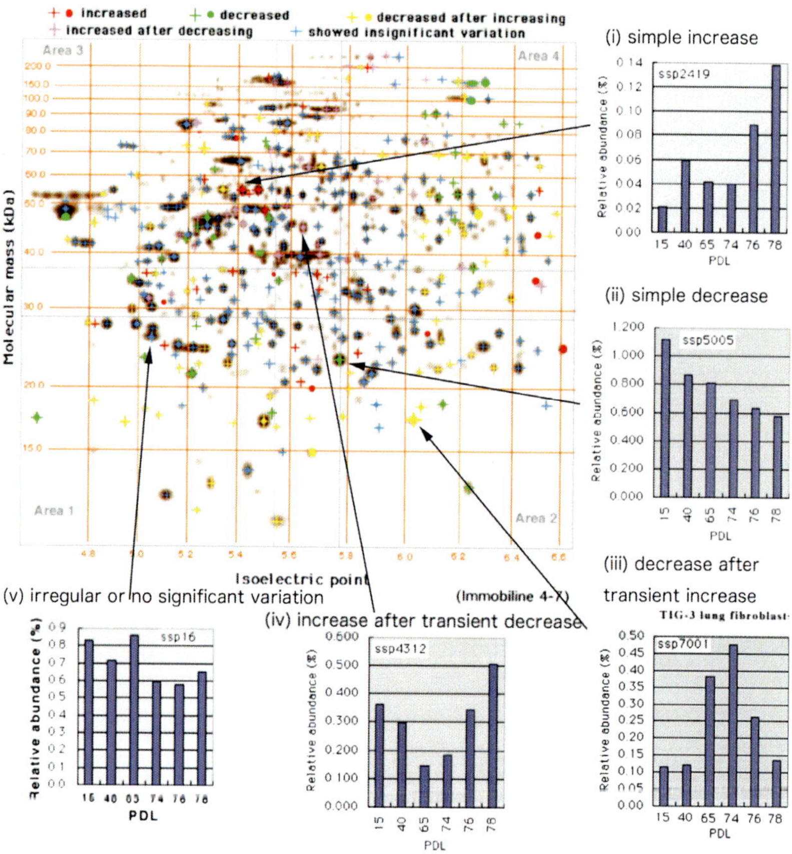

Figure 10.1 Clickable 2-D gel image map of replicative-aging related proteome database.

10.3 Proteome Research on Post-Mitotic Cell Aging in Brain Tissues

Age -related protein alterations appeared in mouse brain tissues have been investigated as a model of post-mitotic cell aging *in vivo* by 2-D gel-based proteome analysis. The brain is situated in the center of the cerebrospinal nervous system in mammals. The brain consists of three main structures: the cerebrum, the cerebellum and the brainstem. Continuing from the brainstem is the spinal cord with cervical, thoracic, lumbar, and sacral regions and peripheral nervous system. We chose the mouse because it is a well-established experimental animal, adequate for preliminary human research. It has served as a model for various human diseases with valuable results already applied to clinical diagnoses and treatments of various diseases. Accordingly, genome analyses have already been being carried out with mice even in relation to human cerebrospinal diseases and aging. We pursue proteome analysis of mouse brain in view of different· anatomic areas, aging, processed and cerebrospinal diseases of normal and genetic traits. Mouse brain proteins were isolated from five regions—cerebellum, cerebral cortex, hippocampus, striatum, and cervical spinal cord—at five ages from the 10th week to the 24th month and separated by 2-DE. Over 1,000 protein spots were visualized by silver staining and quantified by image processing. In the analyses, 58 protein spots were distinguishable among the above five brain regions and 17 proteins varied in quantity in the course of aging. Partial amino-terminal sequences and/or internal sequences for a total of 301 protein spots were analyzed. One hundred and eighty proteins appeared to have blocked N-termini and 122 proteins were identified. Twenty-seven new proteins were identified by sequence homology search. A mouse brain proteome database was constructed, which consisted of the 2-DE map images and the respective spot data files with 15 related references.

10.3.1 Preparation of Tissue Proteins for Analyzing Post-Mitotic Cell Aging

Male mice of C57BL/6 strain at 10 weeks, 3 months, 6 months, 12 months, and 24 months after birth were used. Special care was taken to avoid proteolytic degradation of brain proteins. Cerebellum, cerebral cortex, hippocampus, striatum and spinal cord were quickly removed from the head of the mice after dipping the whole body into liquid nitrogen for 10 sec, near freezing. Each tissue was homogenized with 9.8 M urea, 2% w/v NP-40, 2% v/v Pharmalyte, 100 mM DTT, 0.5 μg/ml E-64, 0.5 mM PMSF, 40 μg/ml TLCK, 1 mg/ml aprotinine, 10 mg/ml chymostatin, 0.5 mM EDTA, and 0.01% w/v bromophenol blue. After incubation at 37°C for 1 h, tissue protein extract for 2-DE was removed by centrifugation. The protein concentration the sample solution was adjusted to 0.14 mg/ml for silver staining, and 1.4 mg/ml for protein identification.

10.3.2 2-D Gel Electrophoresis

The first-dimensional IPG-IEF was carried out on rehydrated DryStrip (18 cm long, non-linear gradient pH3-10) in a horizontal electrophoresis system, at 500 V for 1 h, 1,000 V for 1 h, and finally 8,000 V for 6 h. After the IPG-IEF, the IPG strip was equilibrated at room temperature for 30 min with 50 mM Tris-HCl, pH 6.8, 6 M urea, 3% SDS, 50 mM DTT, and 0.01% BPB. The second-dimensional SDS-PAGE was performed on a 12.5%T running gel (230 × 225 × 1 mm) without the stacking gel.

10.3.3 Visualization of Protein Spots by Staining

Silver stain was carried out using a 2-D silver-stain kit "DAIICHI" II with slight modifications for quantitative image analysis. Proteins were fixed first with 30% methanol, 10% acetic acid containing 0.0025% thiourea for 15 min, then treated 50% methanol, 0.05% DTT, 0.1% glutalaldehyde, and 0.024% thiourea for 10 min prior to treatment with ammoniacal silver nitrate solution. Protein spots were visualized in a developer solution containing 0.005% sodium citrate and 0.02% formaldehyde.

The 2-D gel for protein identification was stained with Coomassie Brilliant Blue (CBB)

10.3.4 Quantitative Spot Analysis by Image Processing

2-D gel images on silver/CBB stained gels were acquired with a scanning densitometer. The digitized image data were analyzed using PDQuest software in almost the same procedure performed in the analysis of replicative aging-related proteome as already mentioned in the text.

10.3.5 Expression Profiling of Protein in Mouse Brain with Aging

To investigate aging-related protein variations in the mouse brain during the course of aging, proteins in extract prepared form mouse brain at five decreased just after and 24th month. The hippocampus showed five spots, all of which continuously decreased after the 10th week. The striatum had two spots that continuously increased after the 6th month. The cervical spinal cord showed five spots, two of which decreased just after the 3rd weeks, while another spot increased after the 6th month and counting to increase until the 24th month, and two other spots increased after the 12th month until 24th month. These changes in five regions of the brain are schematically illustrated in Fig. 10.2. Among them, spot 303 was identified as histone H3.

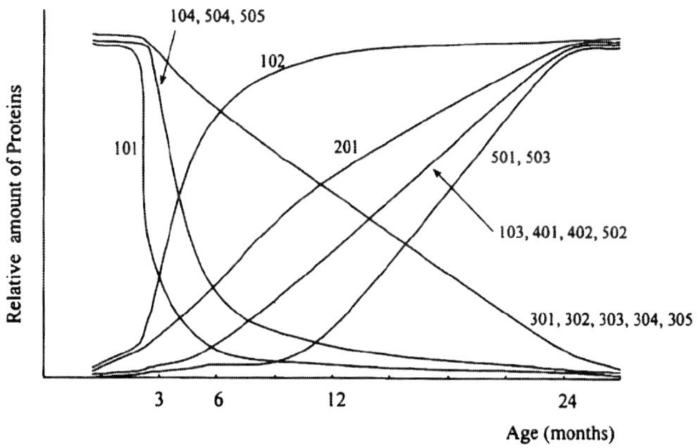

Figure 10.2 Schematic view of aging-related protein variation in mouse brain tissues.

References

1. Hayflick, L., Moorhead, PS. (1961). The serial cultivation of human diploid cell strains. *Exp. Cell Res.*, 25, 585–621.

2. Hayflick, L. (1965). The limited *in vitro* lifetime of human diploid cell strains. *Exp. Cell Res.*, 37, 614–636.
3. Harley CB, Futcher AB, Greider CW. (1990). Telomeres shorten during ageing of human fibroblasts. *Nature*, 345, 458–460.
4. Toda, T., Kaji, K., Kimura, N. (1998). TMIG-2DPAGE: a new concept of two-dimensional gel protein database for research on aging. *Elecrophoresis*, 19, 344–348.
5. Tsugita, A., Kawakami, T., Uchida, T., Sakai, T., Kamo, M., Toda, T. (2000). Proteome analysis of mouse brain: two-dimensional electrophoresis profiles of tissue proteins during the course of aging. *Electrophoresis*, 21, 1853–1871.
6. Kondo, H., Nomaguchi, TA., Sakurai, Y., Yonezawa, Y., Kaji. K., Matsuo, M., Okabe, H. (1988) Effects of serum from human subjects of various ages on proliferation of human lung and skin fibroblasts. *Exp. Cell Res.*, 178, 287–295.
7. Matsuo M, Kaji K, Utakoji T, Hosoda K.((1982). Ploidy of human embryonic fibroblasts during *in vitro* aging. *J. Gerontol.*, 37(1), 33–37.
8. O'Farrell, PH. (1975). High resolution two-dimensional electrophoresis of proteins. *J. Biol. Chem.*, 250, 4007–4021.
9. Görg, A., Boguth, G., Obermaier, C., Posch, A., Weiss, W. (1995). Two-dimensional polyacrylamide gel electrophoresis with immobilized pH gradients in the first dimension (IPG-Dalt): the state of the art and the controversy of vertical versus horizontal systems. *Electrophoresis*, 16, 1079–1086.
10. Celis JE, Dejgaard K, Madsen P, Leffers H, Gesser B, Honore B, Rasmussen HH, Olsen E, Lauridsen JB, Ratz G, *et al.* (1990). The MRC-5 human embryonal lung fibroblast two-dimensional gel cellular protein database: quantitative identification of polypeptides whose relative abundance differs between quiescent, proliferating and SV40 transformed cells. *Electrophoresis*, 11, 1072–1113.
11. Wolters DA, Washburn MP, Yates JR 3rd. (2001). An automated multidimensional protein identification technology for shotgun proteomics. *Anal. Chem.*, 73(23), 5683–5690.
12. MacCoss MJ, McDonald WH, Sara A, Sadiron R, Clark JM, Taste JJ, Gould KL, Wolters D, Washburn M, Weiss A, Clark JI, Yates JR 3rd. (2002). Shotgun identification of protein modifications from protein complexes and lens tissue. *Proc. Natal. Accad. Sci. U S A.*, 99(12), 7900–7905.
13. De St. Groth SF, Webster RG, Datyner A (1963) Two new staining procedures for quantitative estimation of proteins on electrophoretic strips. *Biochem. Biophys. Acta*, 71, 377–391.

14. Congdon RW, Muth GW, Splittgerber AG (1993) The binding interaction of coomassie blue with proteins. *Anal. Biochem.*, 213, 407–413.

15. Switzer RC, Merril CR, Shifrin S (1979) A highly sensitive silver stain for detecting proteins and peptides in polyacrylamide gels. *Anal. Biochem.*, 98, 231–237.

16. Jin LT, Hwang SY, Yoo GS, Choi JK. (2006). A mass spectrometry compatible silver staining method for protein incorporating a new silver sensitizer in sodium dodecyl sulfate-polyacrylamide electrophoresis gels. *Proteomics*, 6(8), 2334–2337.

17. Morrissey JH. (1981). Silver stain for proteins in polyacrylamide gels: a modified procedure with enhanced uniform sensitivity. *Anal. Biochem.*, 117(2), 307–310.

18. Lemkin PF, Lipkin LE. (1981). GELLAB: A computer system for 2D gel electrophoresis analysis I. Segmentation of spots and system preliminaries. *Comput. Biomed. Res.*, 14, 3, 272–297.

19. Anderson NL, Taylor J, Scandora AE, Coulter BP, Anderson NG. 1981. The TYCHO system for computer analysis of two-dimensional gel electrophoresis patterns. *Clin. Chem.* 27(11), 1807–1820.

20. Garrels JI. (1989). The QUEST system for quantitative analysis of two-dimensional gels. *J. Biol. Chem.*, 264(9), 5269–5282.

Chapter 11

Phosphoproteomics of Tumor Cell Lines

Hisafumi Okabe
Research Division, Chugai Pharmaceutical Co., Ltd.,
200 Kajiwara, Kamakura, Kanagawa 247-8530, Japan
okabehsf@chugai-pharm.co.jp

11.1 Introduction

Protein phosphorylation is one of the most significant signal transduction mechanisms by which intercellular signals regulate crucial intracellular processes, and thought to play important roles in the biological aspects in the cells. Protein kinases phosphorylate their substrate proteins, and also can be regulated by activator proteins, inhibitor proteins, ligand binding to regulatory subunits, cofactors, and phosphorylation by other proteins or by themselves. Kinase inhibitors regulate phosphorylation in the network of proteins and make different phosphorylation status in the pathway, which potentially distinguishes the biological output of the cells. Also, kinase inhibitor drugs are used as an important class of new targeted therapeutics in oncology field with the clinical success of several kinase inhibitors including Gleevec and Sutent. The analysis

Quantitative Proteome Analysis: Methods and Applications
Edited by Kazuhiro Imai and Sam F. Y. Li

ISBN 978-981-4316-51-4 (Hardcover), 978-981-4316-98-9 (eBook)
www.panstanford.com

of the phosphorylation status of the tumor cells would give the fruitful information to know mechanism of the tumor cell growth and proliferation. However, detailed molecular understanding of protein phosphorylation is limited. Previously, we performed phosphoproteomic analysis in the presence or absence of nocodazole, a known microtubule-interacting agent, in three distinct tumor cell lines (HeLa, HCT-116, and NCI-H460) and found that 14 proteins were phosphorylated in all three tumor cell lines only as a result of treatment with nocodazole [1]. In this session, we would like to explain the example of phosphoproteomic analysis with tumor cells and the application for biomarker candidates.

11.2 Proteomic Analysis of Phosphorylated Peptides

Proteomic analysis was carried with liquid chromatography–tandem mass spectrometry (LC-MS/MS) after enzymatic digestion of the proteins to digested peptides in the bottom-up proteomics. Especially, the enrichment of phosphorylated peptides enables to analyze phosphopeptides with LC-MS/MS in this decade. Titanium dioxide, immobilized metal affinity chromatography (IMAC), and other metal-based affinity chromatography were widely used to enrich phosphorylated peptides. [2–8] The cells were lysed by sonication and the proteins digested with trypsin followed by enrichment of the phosphorylated peptides with these affinity chromatography. In our laboratory, the lysate was applied on MassPREP™ (Waters, Ma), which is aluminum-based affinity chromatography, according to the manufacturer's protocol. On the analysis of phosphopeptides in mass spectrometer, neutral loss, which is the prior dissociation of phosphates from the parent mass in collision-induced dissociation, is frequently observed. A pseudo MS^3 scan, also referred to as "multiple activation," enables the analysis of neutral loss peptides and enhances detection [9]. Combined with recent advances in mass spectrometry, these technological advances have allowed the identification of hundreds to thousands of phosphorylation sites in a proteome [10–17].

11.3 Phosphoproteomic Analyses of Three Distinct Tumor Cell Lines

Two independent analyses of three distinct tumor cell lines (HeLa, HCT-116, and NCI-H460) with or without nocodazole treatment were performed. The number of identified phosphopeptides was significantly lower in HeLa cells (237 phosphopeptides assigned to 169 phosphoproteins), compared with HCT-116 (648 phosphopeptides assigned to 361 phosphoproteins) or NCI-H460 (664 phosphopeptides assigned to 393 phosphoproteins) (Fig. 11.1). When cells were treated with nocodazole, the number of identified phosphopeptides increased remarkably in HeLa cells (600 phosphopeptides assigned to 359 phosphoproteins) and slightly in NCI-H460 cells (712 phosphopeptides assigned to 405 phosphoproteins) but decreased in HCT-116 cells (400 phosphopeptides assigned to 250 phosphoproteins) (Fig. 11.1). Thus, phosphorylation status and changes in response to nocodazole differed among cell lines. Of the 152 phosphoproteins that were common to all three cell lines with nocodazole, 13 were not identified from any of the cell lines without nocodazole. In contrast, there were no common phosphoproteins present only without nocodazole treatment.

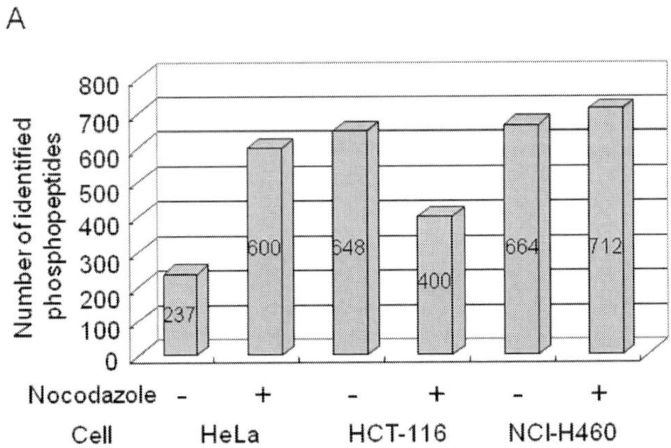

Figure 11.1 Phosphoproteomes of HeLa, HCT-116, and NCI-H460 cells with or without nocodazole treatment. Number of phosphopeptides identified in three tumor cell lines with or without nocodazole treatment. Each sample was analyzed twice.

In total, 1,570 unique phosphopeptides were identified from the analyses and 726 phosphoproteins with unique gene identification (gi) numbers were assigned; 1,196 (76%) of mono-phosphorylated, 349 (22%) of di-phosphorylated, and 25 (2%) of tri- and more phosphorylated peptides were identified, and then, 1,525 phosphorylation sites were determined in this study (Fig. 11.2A); 1,472 (96.5%) of phosphorylated sites were unambiguously assigned exact sites in the peptide sequences from manual inspection, whereas the remaining phosphorylation events mapped to a few possible serine, threonine, or tyrosine residues in peptide sequences. The distribution of phosphorylation sites was 88%, 11%, and 1% for phosphoserine, phosphothreonine, and phosphotyrosine, respectively. Thus, the majority of the identified phosphoproteins were either phosphoserine or phosphothreonine (Fig. 11.2B). This is consistent with the reported phosphoamino acid content ratio (pSer:pThr:pTyr) of a vertebrate cell of 1800:200:1 [18].

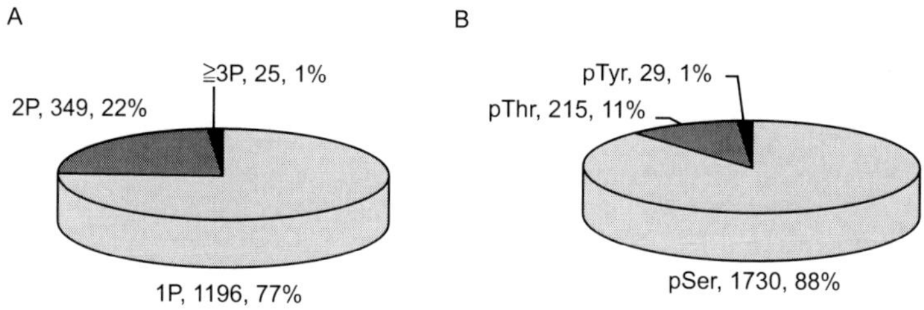

Figure 11.2 Characterization of the identified phosphopeptides: (A) Number of phosphorylation sites on each phosphopeptide. (B) Ratio of phosphorylated amino acid: serine, threonine, and tyrosine.

11.4 Functional Annotation of the Identified Phosphoproteins

The biological processes and molecular functions of identified proteins were annotated according to the ingenuity pathway analysis (IPA; 693 proteins (95%)) among the 726 identified phosphoproteins (Fig. 11.3). The IPA subset identified the numbers of proteins and their biological function, including transcription regulators (64), translation regulators (14), kinases (36), phosphatases (9), peptidases (13), other enzymes (89), transporters (33), ion channels (3), transmembrane receptor/G-protein coupled receptors (4), cytokine/growth factors (3), a nuclear receptor (1), and others. And the IPA subset included the numbers of proteins involved in various biological processes including cell death (138), cellular growth and proliferation (112), gene expression (111), cell cycle (100), cell morphology (81), DNA replication, recombination, and repair (71), cell signaling/cell-to-cell signaling (65), RNA post-transcriptional modification (58), cellular movement (43), cellular development (36), molecular transport (32), nucleic acid metabolism (27), protein synthesis (26), protein trafficking (19), lipid metabolism (10), energy production (10), RNA trafficking (9), post-translational modification (8), RNA damage and repair (1), and others (Fig. 11.4A,B).

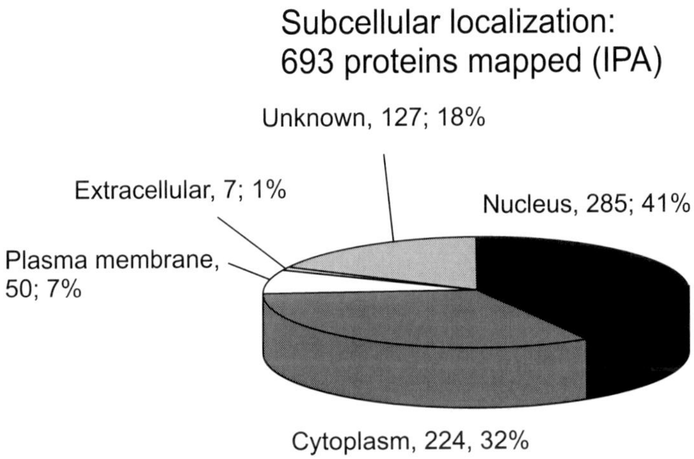

Figure 11.3 Subcellular distribution of the identified phosphoproteome in all three tumor cells classified according to IPA.

In comparison of the functional annotations of each phosphoproteome set obtained from the different cell lines with or without nocodazole treatment, the percentage of proteins in each phosphoproteome in most of the annotation groups was relative to the total number of phosphopeptides but varied among cell lines and in the presence or absence of nocodazole (Fig. 11.5A). However, some functional groups showed changes characteristic to nocodazole-treated cells in all three cell lines. The number of proteins identified with cell morphology, nucleic acid metabolism, and energy production

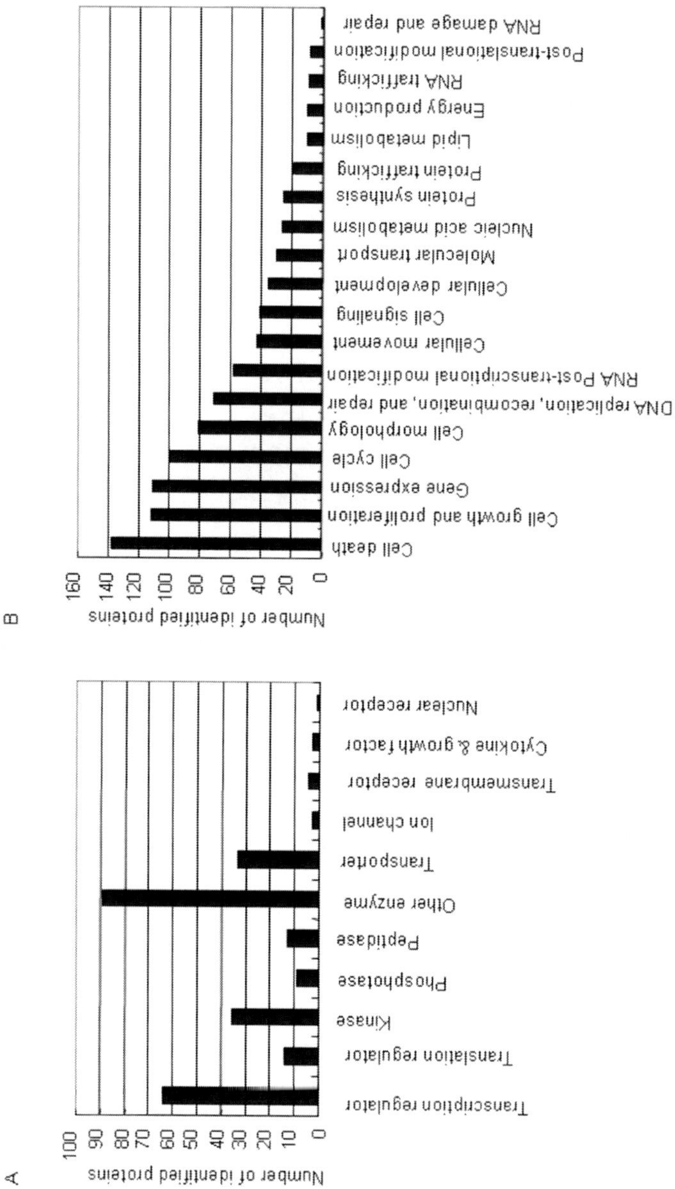

Figure 11.4 Functional annotation of the identified phosphoproteome in all three tumor cells classified according to molecular function (A) and physiological process (B) from IPA.

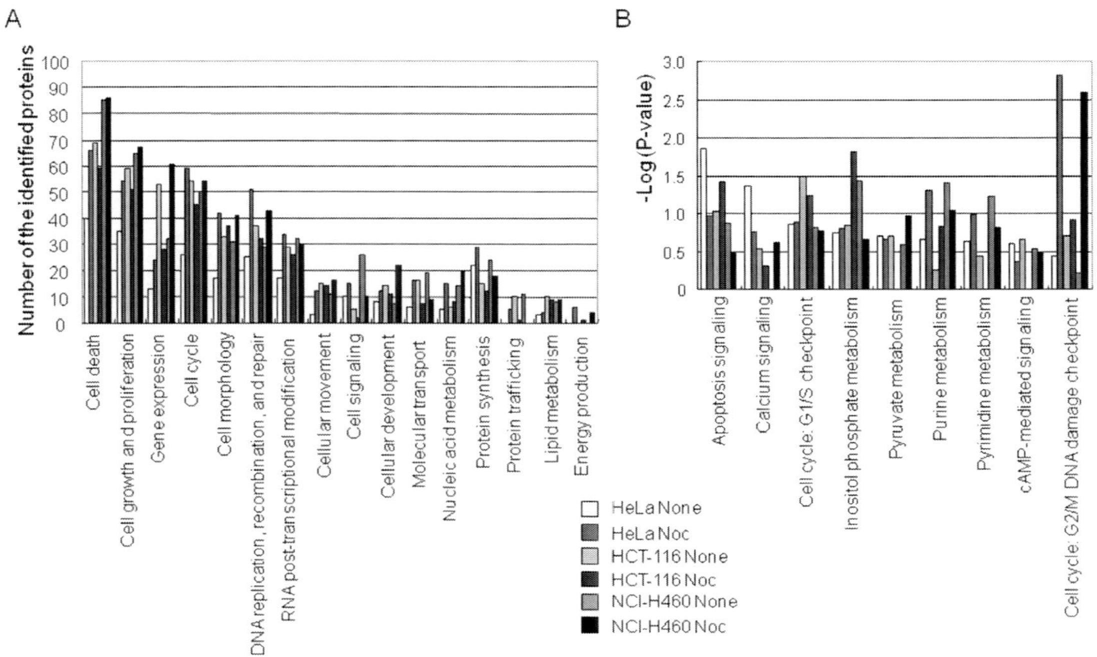

Figure 11.5 Comparison of the number of phosphoproteins involved in the indicated IPA functional annotation of the physiological process (A) and signaling pathway (B) among three tumor cell lines. –Log (*P*-value) on the y-axis in (B) was obtained from the IPA analysis.

showed an increase in all of the nocodazole-treated cell lines on the IPA analysis (Fig. 11.5A). In addition, the cell cycle G2/M checkpoint was remarkable in all nocodazole-treated cells, although instances of other functions such as apoptosis signaling, calcium signaling, cell cycle G1/S checkpoint, inositol phosphate metabolism, and cAMP-mediated signaling differed among the cell lines (Fig. 11.5B).

11.5 Phosphorylation Status of Signaling Proteins and Representative Proteins and Changes upon Nocodazole Treatment in Three Tumor Cell Lines

The phosphopeptides of the signaling proteins include PDPK1, PTK2, CRK, SRC, FYN, PKA, elongation factor-2 kinase, EphA2, GSK3, IGF2 receptor, Erk2, MEK1, Pak2, Src, FAK, STK39, STK4, Crk, Raf, Yes, and WASP. The phosphorylation of PDPK1 S241 is known to be an autophosphorylation site and to induce enzymatic activation, and HNRPK (heterogeneous nuclear ribonucleoprotein K, hnRNP-K) S216, S284, and Y380 are known to be phosphorylated by c-Jun N-terminal kinase (JNK), PKC δ, and Src, respectively. Furthermore, MAPK1 (extracellular signal-regulated kinase 2, Erk2, also called MAPK) T185 and Y187, SRC S75, ARAF (Raf) S582, and WASL Y256 are known to be phosphorylated by MEK1, CDK, AMP-activated protein kinase/PKA, and FAK, respectively (Table 11.1). Especially, the phosphorylation of HNRPK S216 increased with nocodazole treatment in all three cell lines (Table 11.1), suggesting that JNK was activated upon nocodazole treatment. However, the phosphorylation of Src S75 was identified only in nocodazole-treated NCI-H460 cells, and the phosphorylation of STK39 (serine threonine kinase 39) S385 was identified only in non-treated HCT-116 cells (Table 11.1). The phosphorylation of HNRPK S284 was most significant in nocodazole-treated HeLa cells compared with the other cell lines (Table 11.1). It should be noted that those differences may have possibly derived from variations in the LC-MS analyses; even so, the results suggest that the activation state of the signal cascades and the effect of nocodazole were distinctive among cell lines. Also, the phosphorylation status and changes with nocodazole treatment of representative M-phase proteins (Table 11.1) annotated to either

Table 11.1 Examples of signaling molecules and mitotic proteins

Molecule	Site	HeLa Nontreat	HCT-116 Nontreat	NCI-H460 Nontreat	HeLa Noc	HCT-116 Noc	NCI-H460 Noc	Kinase*
Signaling molecules: IPA								
PDPK1	S241	0	0	2	0	0	1	Autophosphorylation
HNRPK	S216	0	1	0	3	3	3	JNK
HNRPK	S284	1	2	2	5	2	2	PKCδ
HNRPK	Y380	0	0	0	1	0	0	Src
MAPK1	T185,Y187	0	0	0	1	0	0	MEK1
MAP2K2	T394	1	1	1	0	1	0	
SRC	S75	0	0	0	0	0	3	CDK group
ARAF	S582	0	0	1	0	0	0	AMPK, PKA
WASL (includes EG:8976)	Y256	0	0	0	1	0	0	FAK
STK39	S385	0	4	0	0	0	0	
Mitosis: GO								
NUMA1	S1757	2	3	2	1	2	1	
TUBA1A	S48	0	0	1	2	0	2	
G2/M transition: IPA								
CDC2	Y15	2	1	1	2	1	0	
CDC2	T14	1	0	2	0	0	2	

(*Continued*)

Table 11.1 *(Continued)*

Molecule	Site	HeLa Nontreat	HCT-116 Nontreat	NCI-H460 Nontreat	HeLa Noc	HCT-116 Noc	NCI-H460 Noc	Kinase*
SFN	S248	0	0	0	1	1	3	
Interested molecules								
KIF21A	S1649	0	1	2	0	0	0	
LMNA	S636	0	1	0	0	2	2	
LMNB1	S23	0	0	1	0	1	2	
LMNB1	T20,S23	0	0	0	1	0	1	
LMNB2	S17	0	1	0	0	0	1	
LMNB2	T14,S17	0	0	0	1	1	0	
LMNB2	T19	0	0	0	0	0	1	
MAP4	S280	1	5	5	4	4	6	
MAPT	S144	0	1	2	0	1	0	
MAPT	S315	0	2	2	0	2	1	

Values are the sum of the number of sequenced MS/MS spectra assigned to the phosphopeptide in 2 independent analyses.

*Known kinases for the identified phosphorylation sites were searched using the Phospho.ELM database (http://phospho.elm.eu.org/)

Of the proteins annotated to "signaling", mitosis, or G2/M transition, examples of the differentially regulated proteins among cell lines are shown in the table.

mitosis by GO or G2/M transition by IPA, and some interested proteins such as microtubule-associated proteins (MAPs), kinesin, and lamins were examined. The phosphorylation of lamin B1 and B2 (LMNB1 and LMNB2), including T20 and S23, and T14, S17, and T19, respectively, was found to be more frequent in the nocodazole-treated cells compared with the cells without nocodazole treatment (Table 11.1). Likewise, the phosphorylation of stratifin (SFN) S248 was found only in the nocodazole-treated cell lines (Table 11.1). In contrast, the phosphorylation of kinesin family member 21A (KIF21A) S1649 was found in the cells without nocodazole treatment but not in the nocodazole-treated cells (Table 11.1). Although there were a few examples of protein phosphorylation with similar changes in all three nocodazole-treated cell lines, most of the mitotic protein phosphorylation with nocodazole treatment differed among the cell lines (Table 11.1).

11.6 Nocodazole-Induced Phosphoproteins Common to Three Distinct Tumor Cell Lines

The activation state of signaling molecules and regulation of M-phase proteins were distinctive among the cell lines. Nevertheless, some proteins were phosphorylated in all three tumor cell lines only as a result of treatment with nocodazole (Table 11.2). The phosphorylation of nucleophosmin 1 isoform 2 (NPM1) S225 corresponding to nucleophosmin 1 isoform 1 S254 and coatomer protein complex, subunit alpha isoform 1 (COPA) S173, were identified in all independent LC-MS analyses of the nocodazole-treated cells (Table 11.2).

11.7 A Time Course Analysis of Nocodazole-Induced Phosphorylations

Protein phosphorylation has been shown to affect mitotic events such as mitotic checkpoints, spindle formation, and the anaphase-promoting complex. Then, the phosphoproteomic analysis of HCT-116 cells treated with nocodazole, which induces the activation of stress response pathways, M-phase cell cycle arrest, and the induction

Table 11.2 Nocodazole-induced phosphorylation

Molecule	Sequence*	Site	HeLa Nontreat	HCT-116 Nontreat	NCI-H460 Nontreat	HeLa Noc	HCT-116 Noc	NCI-H460 Noc
NPM1	MQA**S**IEKGGSLPK	S254	0	0	0	3	3	5
COPA	NL**S**PGAVESDVR	S173	0	0	0	2	2	2
PSMA5	GVNTF**S**PEGR	S16	0	0	0	2	2	1
ILF3	LFPD**T**PLALDANK	T592	0	0	0	2	1	4
PCBP1	VMTIPYQPMPA**S**SPVICAGGQDR	S189	0	0	0	2	1	3
EIF2A	SDK**S**PDLAPTPAPQSTPR	S506	0	0	0	1	2	1
MYO1C	DVE**S**PSWR	S389	0	0	0	1	1	2
RPL12	IGPLGL**S**PK	S38	0	0	0	1	1	2
TK1	KLFAPQQILQC**S**PAN	S231	0	0	0	1	1	2
TMPO	FQETEFL**S**PPR	S424	0	0	0	1	1	4
SFN	DNLTLWTADNAGEEGGEAPQEPQ**S**	S248	0	0	0	1	1	3
RBM4	LHVGNI**S**PTCTNK	S86	0	0	0	1	1	2
RIC8A	VIQPMGM**S**PR	S508	0	0	0	1	1	1
IARS	APLKPYPV**S**PSDK	S1047	0	0	0	1	1	1

Values are the sum of the number of sequenced MS/MS spectra assigned to the phosphopeptide in 2 independent analyses.

*Identified phosphorylation sites are shown in boldface type.

of apoptosis for 0, 6, 10, 14, or 24 h containing <10%, ~30%, ~50%, ~80%, or >90% M-phase cells, respectively, is performed. Temporal changes of phosphorylation induced by nocodazole varied among the molecules. Phosphorylations of NPM1 S254, PSMA5 S16, ILF3 T592, and TMPO S424 were induced by nocodazole after 6 h and sustained until 24 h. This result implies that the phosphorylation of NPM1 S254 may be the early indicator of nocodazole treatment and may also be functionally involved in the induction of M-phase arrest by nocodazole. Phosphorylations of COPA S173, TK1 S231, and RIC8A S508 were observed after 14 h or more when the majority of the cells were arrested in M-phase. Especially, the phosphorylation of COPA S173 was observed from 14 h of nocodazole treatment when ~80% of cells were in M-phase, suggesting that COPA S173 may be phosphorylated as a result of M-phase arrest and may also be a candidate for an M-phase-specific biomarker. Phosphorylations of MYO1C S389 and SFN S248, which were identified only from cells treated with nocodazole for 10 h and 14 h, respectively, seemed to be induced transiently in the course of the nocodazole treatment. Our results suggest that phosphorylation of NPM1 S254 and COPA S173 may serve as biomarkers to monitor the efficacy of microtubule-interfering agents for cancer chemotherapy.

11.8 Conclusion

Protein phosphorylation is thought to play an important role in a wide variety of the biological activities of a drug. The phosphorylation status in the pathway potentially determines a distinct biological output. We performed phosphoproteomic analysis of three distinct tumor cell lines (HeLa, HCT-116, and NCI-H460) and also a time course analysis of nocodazole-induced phosphorylations of HCT-116 cells with nocodazole treatments. The results suggest potential candidates for biomarkers commonly induced by nocodazole in distinct tumor cell lines. Further study will be required to validate whether the changes observed are consistent with other types of tumor cell lines and/or clinical samples and to elucidate the function of these phosphorylations in mitosis and from nocodazole treatment. Our approach may also have clinical implications for optimizing cancer therapies involving microtubule-interfering drugs in future.

References

1. Nagano, K., Shinkawa, T., Mutoh, H., Kondoh, O., Morimoto, S., Inomata, N., Ashihara, M., Ishii, N., Aoki, Y., and Haramura, M. (2009). Phosphoproteomic analysis of distinct tumor cell lines in response to nocodazole treatment, *Proteomics*, **9**, pp. 2861–2874.
2. Nuwaysir, L. M., and Stults, J. T. (1993). Electrospray ionization mass spectrometry of phosphopeptides isolated by on-line immobilized metal-ion affinity chromatography, *J. Am. Soc. Mass Spectrom.*, **4**, pp. 662 –667.
3. Xhou, W., Merrick, B. A., Khaledi, M. G., and Tomer, K. B. (2000) Detection and sequencing of phosphopeptides affinity bound to immobilized metal ion beads by matrix-assisted laser desorption/ionization mass spectrometry, *J. Am. Soc. Mass Spectrom*, **11**, pp. 273–282.
4. Muszynska, G., Dobrowolska, G., Medin, A., Ekman, P., and Porath, J. O. (1992) Model studies on iron(III) ion affinity chromatography. II. Interaction of immobilized iron(III) ions with phosphorylated amino acids, peptides and proteins, *J. Chromatogr.*, **604**, pp. 19–28.
5. Posewitz, M. C., and Tempst, P. (1999) Immobilized gallium(III) affinity chromatography of phosphopeptides, *Anal. Chem.*, **71**, pp. 2883 –2892.
6. Andersson, L., and Porath, J. (1986) Isolation of phosphoproteins by immobilized metal (Fe^{3+}) affinity chromatography. *Anal. Biochem.*, **154**, pp. 250–254.
7. Pinkse, M. W., Uitto, P. M., Hilhorst, M. J., Ooms, B., and Heck, A. J. (2006) Selective isolation at the femtomole level of phosphopeptides from proteolytic digests using 2D-NanoLC-ESI-MS/MS and titanium oxide precolumns, *Anal. Chem.*, **76**, pp. 3935 –3943.
8. Larsen, M. R., Thingholm, T. E., Jensen, O. N., Roepstorff, P., and Jorgensen, T. J. (2005) Highly selective enrichment of phosphorylated peptides from peptide mixtures using titanium dioxide microcolumns, *Mol. Cell. Proteomics.*, **4**, pp. 873 –886.
9. Schroeder, M. J., Shabanowitz, J., Schwartz, J. C., Hunt, D. F., and Coon, J. J., (2004) A neutral loss activation method for improved phosphopeptide sequence analysis by quadrupole ion trap mass spectrometry, *Anal. Chem.* **76**, pp. 3590–3598.

10. Kim, J. E., Tannenbaum, S. R., and White, F. M. (2005) Global phosphoproteome of HT-29 human colon adenocarcinoma cells, *J. Proteome Res.*, **4**, pp. 1339 –1346.

11. Yang, F., Stenoien, D. L., Strittmatter, E. F., Wang, J., Ding, L., *et al.* (2006) Phosphoproteome profiling of human skin fibroblast cells in response to low and high-dose irradiation, *J. Proteome Res.*, **5**, pp. 1252 –1260.

12. Ballif, B. A., Villen, J., Beausoleil, S. A., Schwartz, D., and Gygi, S. P. (2004) Phosphoproteomic analysis of the developing mouse brain., *Mol. Cell. Proteomics.*, **3**, pp. 1093–1101.

13. Beausoleil, S. A., Jedrychowski, M., Schwartz, D., Elias, J. E., Villen, J., *et al.* (2004) Large-scale characterization of HeLa cell nuclear phosphoproteins, *Proc. Natl. Acad. Sci. USA.*, **101**, pp. 12130–12135.

14. Ficarro, S. B., McCleland, M. L., Stukenberg, P. T., Burke, D. J., Ross, M. M., *et al.* (2002) Phosphoproteome analysis by mass spectrometry and its application to Saccharomyces cerevisiae, *Nat. Biotechnol.*, **20**, pp. 301–305.

15. Nuhse, T. S., Stensballe, A., Jensen, O. N., and Peck, S. C. (2003) Large-scale analysis of *in vivo* phosphorylated membrane proteins by immobilized metal ion affinity chromatography and mass spectrometry, *Mol. Cell. Proteomics.*, **2**, pp. 1234–1243.

16. Shu, H., Chen, S., Bi, Q., Mumby, M., and Brekken, D. L. (2004) Identification of phosphoproteins and their phosphorylation sites in the WEHI-231 B lymphoma cell line, *Mol. Cell. Proteomics.*, **3**, pp. 279–286.

17. Olsen, J. V., Blagoev, B., Gnad, F., Macek, B., Kumar, C., *et al.* (2006) Global, *in vivo*, and site-specific phosphorylation dynamics in signaling networks, *Cell.*, **127**, pp. 635–648.

18. Mann, M., Ong, S. E., Gronborg, M., Steen, H., Jensen, O. N., and Pandey, A., (2002) Analysis of protein phosphorylation using mass spectrometry: deciphering the phosphoproteome, *Trends. Biotechnol.*, **20**, pp. 261–268.

Chapter 12

HCV Infection and Mitochondria Proteomics

Kyoji Moriya,[a] Takeya Tsutsumi,[b] Hideyuki Miyoshi,[c] and Kazuhiko Koike[c]
[a]*Department of Infection Control and Prevention, Graduate School of Medicine, The University of Tokyo, Japan*
[b]*Department of Infectious Diseases, Graduate School of Medicine, The University of Tokyo, Japan*
[c]*Department of Gastroenterology, Graduate School of Medicine, The University of Tokyo, Japan*
moriya-tky@umin.ac.jp

HCV infects approximately 170 million people, including 2 million in Japan, and it is estimated that 3–4 million people are infected with HCV each year.

HCV induces serious chronic hepatitis that results in the development of steatosis, cirrhosis and ultimately hepatocellular carcinoma (HCC). HCV core protein is well known to be the viral capsid protein as well as the pathogenic factor that induces steatosis and HCC without apparent inflammation in transgenic mice model. The functions of the core on the modulation of cellular events have been extensively examined and characterized. Recently, quantitative

Quantitative Proteome Analysis: Methods and Applications
Edited by Kazuhiro Imai and Sam F. Y. Li

ISBN 978-981-4316-51-4 (Hardcover), 978-981-4316-98-9 (eBook)
www.panstanford.com

mitochondrial proteomics show promising results for exploring the mechanism of HCV pathogenesis at the protein level.

In this review, we summarize the current status of our knowledge regarding the pathogenicity of HCV core protein and the findings from recent proteomic surveys of mitochondrial proteins using HCV core gene transgenic mice.

12.1 HCV

With an estimated 3% of the world's population currently infected with hepatitis C, and approximately 170 million persons at risk of hepatitis disease, the World Health Organization (WHO) recognizes hepatitis C as a global health problem [1]. Many HCV-infected individuals develop chronic hepatitis, which eventually progresses into liver cirrhosis and hepatocellular carcinoma.

The shape of HCV is enveloped in a lipid bilayer in which envelope proteins (Envelope) are anchored. The envelope surrounds the nucleocapsid, which is composed of multiple copies of a small basic protein (core), and contains the RNA genome.

HCV envelope proteins interact with several host receptors. The receptor proteins for HCV entry into cells are identified: CD81, scavenger receptor class B type I or SCARB1, and claudin-1. A fourth cellular protein, occludin, is found to be essential for HCV entry into cells [2]. Recently, liver microRNA miR-122 [3, 4] has been thought to be essential for viral infectivity. Gene expression profiling and proteomic approaches have led to the identification of several host–viral interactions [5].

Detailed analyses of HCV have been hampered by the lack of viral culture systems. Recently, the development of HCV strain JFH-1, which generates infectious HCV in culture, has made a contribution to the study of the HCV life cycle [6].

12.2 HCV Genome

Hepatitis C virus (HCV) is an enveloped RNA virus of the Flavivirus family, in which a positive-sense, single-stranded RNA genome of approximately 9600 nucleotides (nt) is contained within the nucleocapsid (Fig. 12.1). The genome consists of a large translational open reading frame (ORF) encoding a polyprotein of approximately

3010 amino acids (aa) . The ORF is flanked by highly conserved untranslated regions (UTR) at both the 5′ and 3′ termini. The complete 5′ UTR consists of 341 nt and acts as an internal ribosomal entry site. This feature leads to the translation of the RNA genome using a cap-independent mechanism, rather than ribosome scanning from the 5′ end of a capped molecule. The polyprotein is processed by both the cellular and viral proteases to generate the viral gene products, which are subdivided into the structural and non-structural proteins. The structural proteins, which are encoded by the NH2-terminal quarter of the genome, include the core protein and the envelope proteins, E1 and E2. The E2 has an alternative form, E2-p7, though it is not clear whether the p7 composes the viral particle. The NS2, NS3, NS4A, NS4B, NS5A, and NS5B are the non-structural proteins that are coded in the remaining portion of the polyprotein. These include serine protease (NS3/4A), NTPase/helicase (NS3) and RNA-dependent RNA polymerase (NS5B).

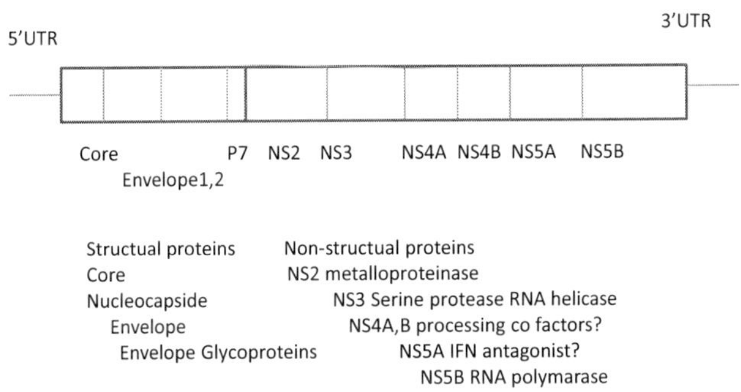

Figure 12.1 Hepatitis C virus structure. Map of the HCV genome. The 9500-nucleotide-long positive-strand RNA of HCV is shown. It encodes a single 3000-amino-acid polyprotein that is proteolytically cleaved into mature proteins by virally encoded proteases.

12.3 HCV Core Protein

The core protein of HCV occupies residues 1–191 of the precursor polyprotein and is cleaved between the core and E1 protein by host signal peptidase. The C-terminal membrane anchor of the core protein is further processed by host signal peptide peptidase [7]. The mature core protein is estimated to consist of 177–179 amino acids and shares high homology among HCV genotypes. The HCV core protein possesses the hydrophilic N-terminal region "domain 1" (residues 1–117) followed by a hydrophobic region called "domain 2," which is located from residue 118–170. The domain 1 is rich in arginine and lysine, and is implicated in RNA-binding and homo-oligomerization. The amphipathic helices I and II spanning from residue 119–136 and residue 148–164, respectively, in domain 2 are involved in the association of HCV core protein with lipid [8].

In addition, the region spanning from residue 112–152 is associated with membranes of the endoplasmic reticulum and mitochondria [9]. The core protein is also localized into the nucleus [10, 11] and binds to the nuclear proteasome activator PA28g/REGg, resulting in PA28g-dependent degradation of the core protein [12].

A recent report suggests that ubiquitination and adenosine triphosphate (ATP) are not required for PA28g-dependent proteasome activity [13]. HCV core protein is also known to be ubiquitinated by E3 ligase E6AP and degraded in the ubiquitin/ATP-dependent pathway [10].

In summary, the HCV core protein—unglycosylated protein—is mainly located at the endoplasmic reticulum as well as mitochondria and lipid droplets within the cytoplasm and also detected in the nucleus.

It has been reported to interfere with cell signaling by modulating mitogen-activated protein kinase (MAPK) signaling [14], interacting with STAT3 and RXR and modifying the expression of cellular protooncogenes such as c-myc and tumor suppressor genes (p53, p21, and pRb) [15]. In cooperation with H-ras, the HCV core has been reported to transform both immortalized and primary rat fibroblasts [16]. Our transgenic mice carrying HCV core gene develop HCC [17, 18]. This information implies that the core protein has a direct effect on the pathogenesis of diseases caused by HCV.

12.4 Transgenic Mice Model

We have engineered transgenic mouse lines carrying the HCV genome by introducing the genes from the cDNA of the HCV genome of genotype 1b (Fig. 12.2) [17, 18]. Established are three different kinds of transgenic mouse lines, which carry the core gene, envelope genes or non-structural genes, respectively, under the same transcriptional regulatory element. Among these mouse lines, only the transgenic mice carrying the core gene developed HCC in two independent lineages. The envelope gene transgenic mice do not develop HCC [19, 20]. Interestingly, the envelope gene transgenic mice develop an exocrinopathy in the salivary and lachrymal glands resembling Sjögren syndrome occurring in cases with human chronic HCV infection. The transgenic mice carrying the entire non-structural genes have developed no HCC.

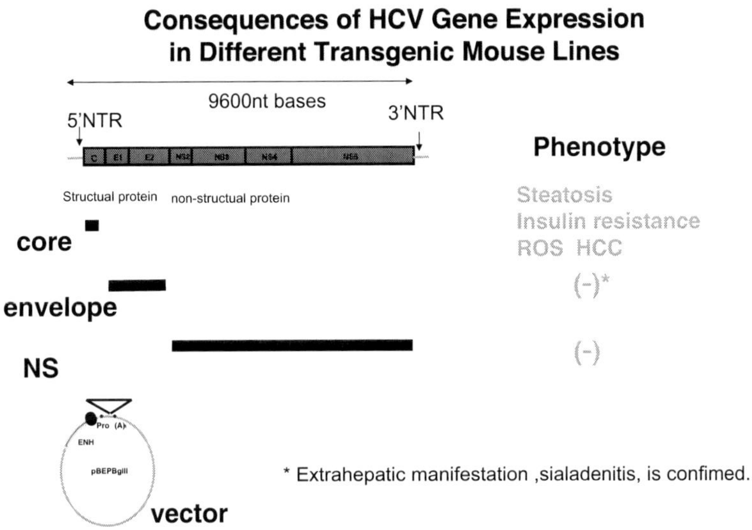

Figure 12.2 HCV structural protein has an effect on pathological phenotypes in the liver. Transgenic mice expressing HCV core protein develop hepatic steatosis and hepatocellular carcinoma (HCC). Transgenic mice that carry HCV envelope genes show no pathological changes, such as hepatitis or hepatic neoplasia, in livers. But these mice develop an exocrinopathy in the salivary and lachrymal glands resembling Sjögren syndrome.

12.5 HCV Core Transgenic Mice

We report the development of HCC in two independent lines of mice transgenic for the HCV core gene, which develop hepatic steatosis early in life as a histological feature characteristic of chronic hepatitis C along with lymphoid follicle formation and bile duct damage [21]. After the age of 16 months, late in life, these transgenic mice develop

HCC. Transgenic mice of both lines developed hepatic tumors that first appeared as adenomas containing fat droplets in the cytoplasm. Then HCC, a more poorly differentiated neoplasia, developed from within the adenomas, presenting in a "nodule-in-nodule" manner without cytoplasmic fat droplets; this closely resembled the histopathological characteristics of the early stage of HCC in patients with chronic hepatitis C. HCV transgenic mouse models also show a disappearance of the double structure of mitochondrial membranes in HCV core transgenic mice livers, and this suggests the interaction between HCV core protein and mitochondria.

12.6 HCV Infection and Metabolism

As described before, steatosis is often observed in chronic hepatitis C patients and is significantly associated with increased fibrosis and progression rate of fibrosis of the liver [22]. A comprehensive analysis of gene expression in the liver of core gene transgenic mice, in which steatosis develops from early in life, revealed that a number of genes related to lipid metabolism are significantly up- or down-regulated.

The composition of fatty acids that are accumulated in the liver of core gene transgenic mice is different from that in fatty liver due to simple obesity. Carbon 18 mono-unsaturated fatty acids (C18:1) such as oleic or vaccenic acids are significantly increased. This is also the case in the comparison of liver tissues from hepatitis C patients and simple fatty liver patients due to obesity. The mechanism of steatogenesis in hepatitis C was investigated using this mouse model [23].

There are several pathways for the development of steatosis. One is the frequent presence of insulin resistance in hepatitis C patients as well as in the core gene transgenic mice, which occurs through the inhibition of tyrosine-phosphorylation of insulin receptor substrate (IRS)-1.26 [24]. Insulin resistance increases the peripheral release and hepatic uptake of fatty acids, resulting in an accumulation of lipid in the liver. In the second one, the HCV core protein targets microsomal triglyceride transfer protein (MTP) activity, thus interfering with the hepatic assembly and secretion of apolipoprotein (apo) B-containing very-low-density lipoproteins (VLDL) [25]. This inhibits the secretion of very-low-density protein (VLDL) from the liver, yielding an increase of triglycerides in the liver. The third one involves the sterol regulatory element-binding protein (SREBP)-1c, which regulates the production of triglycerides and phospholipids. In HCV core gene transgenic mice, SREBP-1c is activated, while neither SREBP-2 nor SREBP-1a is unregulated [26]. This corroborates the results of in vitro studies [27, 28] and a chimpanzee study. Also, the core protein is shown to bind with retinoid X receptor (RXR)-α, resulting in the up-regulation of some lipid metabolism enzymes, including cellular retinol binding protein II and acyl-CoA oxidase [29]. In addition, the persistent activation of peroxisome proliferator activated receptor (PPAR)-α has recently been found in the liver of HCV core gene transgenic mice, yielding changes in lipid metabolism and hepatocyte proliferation, including HCC development [30, 31].

12.7 Steatosis and Mitochondria

There is also a marked accumulation of triglycerides in core-expressing HepG2 cells. While the oleic/stearic acid (18:1/18:0) and palmitoleic/palmitic acid ratio (16:1/16:0) are comparable in both the core-expressing HepG2cells and the control cells, there is a marked accumulation of downstream product, 5,8,11-eicosatrienoic acid (20:3(n-9)) in the core-expressing HepG2 cells. The addition of eicosatetraynoic acid, which inhibits delta-6 desaturase activity which is inherently high in HepG2 cells, led to a marked accumulation of oleic and palmitoleic acids in the core-expressing cells, showing that delta-9 desaturase is activated by the core protein. Eicosapentaenoic acid (20:5(n-3)) or arachidonic acid (20:4(n-6)) administration significantly decreases delta-9 desaturase activity, the concentration of 20:3(n-9), and triglyceride accumulation. This lipid metabolism disorder is associated with NADH accumulation due to mitochondrial dysfunction and is reversed by the addition of pyruvate through NADH utilization [32].

Delta-9 desaturase, stearoyl-CoA desaturase (SCD) is the rate-limiting enzyme in the biosynthesis of monounsaturated fatty acids. It catalyzes the introduction of the cis double bond in the delta-9 position of fatty acyl-CoA substrates. The major monounsaturated fatty acids of triglyceride (TG), cholesterol esters, and membrane phospholipids are palmitoleic and oleic acids. The ratio of stearic acid to oleic acid is one of the factors influencing cell membrane fluidity.

It is known that a deficiency of SCD-1 ameliorates the obesity of ob/ob mice and completely corrects the hypometabolic phenotype of leptin deficiency. These findings suggest that down-regulation of SCD-1 is an important component of the metabolic response and suggests that inhibition of SCD-1 could be of benefit for the treatment of hepatic steatosis, and even other metabolic disorders [33].

12.8 Virus Infection and Mitochondria Function

Many viruses have evolved mechanisms to alter mitochondrial function. The mechanisms are varied, and mitochondria are affected by both direct interactions with viral proteins and by secondary effects of viral-activated signaling cascades with suppression or induction of apoptosis and enhanced generation of reactive oxygen species (ROS).

The HCV produces a viral core protein that targets the mitochondria and increases ROS production.

Oxidative stress overproduction with the suppression of some antioxidative molecules and intracellular signaling augmentation are shown to be the key events in HCV-associated hepatocarcinogenesis.

12.9 HCV and Reactive Oxygen Species

There is a notable feature in the localization of the core protein in hepatocytes: While the core protein predominantly exists in the cytoplasm associated with lipid droplets, it is also present in the mitochondria and nuclei (Fig. 12.3). On the basis of this finding, the pathways related to the mitochondria and nuclei were investigated.

We examined several parameters of oxidative stress and redox homeostasis in a mouse model of HCV-associated HCC. For young mice ages 3–12 months, there is no significant difference in the levels of hydroperoxides of phosphatidylcholine (PCOOH) and phosphatidylcthanolamine in liver tissue homogenates between transgenic and nontransgenic control mice. In contrast, the PCOOH level is increased in old core gene transgenic mice. Concurrently, there is a significant increase in the catalase activity, and there are decreases in the levels of antioxidants, total and reduced glutathione in the same mice. A direct in situ determination by chemiluminescence reveals an increase in hydroperoxide products even in young transgenic mice, suggesting that hydroperoxides is overproduced but immediately removed by an activated scavenger system in young mice. Electron microscopy reveals lipofuscin granules, secondary lysosomes carrying various cytoplasmic organelles, and disruption of the double membrane structure of mitochondria, and PCR analysis discloses a deletion in mitochondrial DNA. Interestingly, alcohol causes a marked increase in the PCOOH level in transgenic mice, suggesting synergism between alcohol and HCV in hepatocarcinogenesis [34]. The HCV core protein alters the oxidant/antioxidant state in the liver in the absence of inflammation and may thereby contribute to or facilitate, at least in part, the development of HCC in HCV infection. In addition, analysis of the antioxidant system revealed that some antioxidative molecules are not increased despite the overproduction of ROS in the liver of core gene transgenic mice [34].

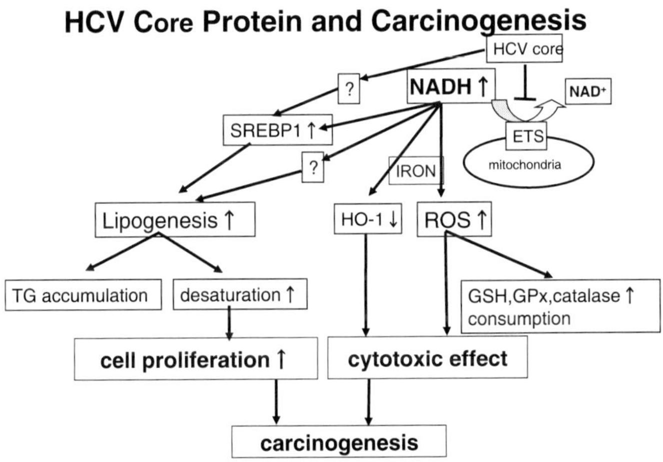

Figure 12.3 HCV core protein has multiple molecular mechanisms that induce the development of pathogenic processes. HCV core protein induces alterations in cellular redox state (decrease in the NAD+/NADH ratio), through effect on mitochondria ETS and increases reactive oxygen species (ROS) production. Excess iron in the liver may promote the generation of ROS. HCV core gene expression is also associated with decreased antioxidant HO-1 expression. This ROS production may, in part, be responsible for the development of HCC. HCV core protein also changes the expression of glucose and lipid metabolism-related genes, thereby causing metabolism disorders.

Iron is an essential cofactor for important biological activities and biochemical reactions, including the transport of oxygen via red blood cells and its reduction to water during respiration. And the regulation of iron homeostasis is tightly controlled. While iron's bioavailability is generally limited, pathological accumulation of the metal within liver aggravates the generation of ROS and elicits toxic effects, which are mainly related to oxidative stress.

One of the characteristics of HCV infection is the unusual augmentation of oxidative stress, which is exacerbated by iron accumulation in the liver, as observed frequently in hepatitis C patients. Using a transgenic mouse model, the core protein of HCV induces the overproduction of ROS in the liver. Iron overloading causes the induction of ROS as well as antioxidants in HCV core transgenic mice liver. However, the augmentation of some antioxidants, including heme oxygenase-1 and NADH dehydrogenase, quinone 1, is compromised by the presence of the core protein. The attenuation of iron-induced augmentation of heme oxygenase-1 is also confirmed in HepG2 cells expressing the core protein. This attenuation is not dependent on the Nrf2 transcription factor. Thus, HCV infection not only induces oxidative stress but also hampers the iron-induced antioxidant activation in the liver [35].

These results suggest that HCV core protein not only induces the overproduction of ROS but also attenuates some of the antioxidant system, which may explain the mechanism underlying the production of a strong oxidative stress in HCV infection compared with other forms of hepatitis. Thus, in the absence of inflammation, the core protein induces oxidative stress overproduction, which may, at least in part, contribute to hepatocarcinogenesis in HCV infection. If inflammation were added to the liver with the HCV core protein, the production of oxidative stress would be escalated to an extent that cannot be scavenged anymore by a physiological antagonistic system. The basis for the overproduction of oxidative stress may be ascribed to the mitochondrial dysfunction [18, 34]. The dysfunction of the electron transfer system of the mitochondrion is suggested in association with the presence of the HCV core protein [36]. Other HCV proteins—core, E1, and NS3—are potent ROS inducers and their expression causes the DNA damage.

Other pathways in hepatocarcinogenesis would be the alteration of the expression of cellular genes and modulation of intracellular signaling pathways. For example, tumor necrosis factor (TNF)-a and interleukin (IL)-1b have been found transcriptionally activated [37]. The mitogen-activated protein kinase (MAPK) cascade is also activated in the liver of the core gene transgenic mouse model [38].

12.10 Mitochondria

HCV core protein has direct effects on mitochondria function, especially on mitochondria electron transport system, and results in ROS production.

Mammalian mitochondria contain over 3000 proteins and an average of 5 copies of their own ring-shaped DNA. Mitochondrial DNA is the only DNA that we inherit exclusively from our mothers. It codes for just 13 mitochondrial proteins. All the other proteins in mitochondria (>99%) come from genes in the nucleus. As a cell makes only about 10,000 to 15,000 proteins, mitochondria contain 20–30% of all the proteins in the cell. Thirteen of them are encoded in mitochondrial DNA and are synthesized inside the mitochondrion. The remainder are encoded in the nucleus, synthesized in the cellular cytoplasm and imported into the organelle.

Mitochondria are of central importance for energy generation. Expression changes or functional alterations in mitochondrial enzymes play a key role.

Quantitative mitochondrial proteomics by using HCV CORE transgenic mice livers show promising results for exploring the mechanism of HCV carcinogenesis at the protein level.

12.11 HCV and Prohibitin

The recent progress in proteomics has opened new avenues for disease-related biomarker discovery. Among proteomics approaches, two-dimensional polyacrylamide gel electrophoresis (2D-PAGE) is a technique for the separation and identification of proteins in a sample by displacement in two dimensions oriented at right angles to one another. This method is generally used as a component of proteomics and is the step used for the isolation of proteins for further characterization by mass spectrometry. 2D-PAGE is particularly useful when comparing two related samples such as healthy and diseased tissue. For example, proteins that are more abundant in diseased tissue may represent novel drug targets or diagnostic markers. In fact, several candidate biomarkers for many human cancers have been identified by this approach. There are, however, tens of thousands of proteins in a cell, differing in abundance over six orders of magnitude. 2D-PAGE is not sensitive enough to detect rare proteins, and hence many proteins are not resolved. Therefore, splitting a sample into different fractions is often necessary to reduce the complexity of protein mixtures prior to 2D-PAGE. For this advantage, Lescuyer *et al.* performed a 2D-PAGE of human mitochondrial proteins derived from the placenta and identified proteins mainly by peptide mass fingerprinting.

Two-dimensional polyacrylamide gel electrophoresis of mitochondria isolated from HepG2 cells reveals that prohibitin, a mitochondrial protein chaperon, is up-regulated not only in core-expressing cells but also in full-genomic replicon cells and livers of core-gene transgenic mice. The stability of prohibitin is increased through interaction with the core protein. The interaction of prohibitin with mitochondrial DNA-encoded subunits of cytochrome c oxidase (COX) is disturbed by the core protein, resulting in a significant decrease in COX activity. The HCV core protein affects the steady-state levels of a subset of mitochondrial proteins including prohibitin, which may lead to an impaired function of the mitochondrial respiratory chain with the overproduction of oxidative stress [39].

12.12 LC-Tandem Mass Spectrometry

A sensitive quantitative method is required for differential profiling analyses of clinical proteomes to understand the disease progress. The FD-LC–MS/MS method, consisting of fluorogenic derivatization (FD), separation by liquid chromatography (LC), and identification by LC-tandem mass spectrometry (MS/MS) is notable for its high resolution, sensitivity, and reproducibility.

Ichibangase shows that disease-related proteins in livers of HCV core gene transgenic (Tg) and non-transgenic (NTg) mice at three developmental stages. After 6 months, the expression of apoptosis-related proteins is suppressed. After 12 months, proteins related to respiration, the electron-transfer system, and anti-oxidation are significantly up-regulated. After 16 months, proteins related to defense, β-oxidation, and apoptosis are significantly suppressed. This fluctuating expression of proteins related to mitochondria function could explain the progression of hepatocarcinogenesis [40].

12.13 Immunity

Viral infection results in the activation of multiple signaling pathways, but recent studies implicate a new role of mitochondria in antiviral innate immunity. The retinoid-inducible gene 1 (RIG-I) is a cytosolic pathogen recognition receptor that engages viral RNA in infected cells to trigger innate immune defenses through its adaptor protein MAVS. MAVS resides on mitochondria and peroxisomes, and a major site of MAVS signaling is the mitochondrial-associated membrane (MAM), a distinct membrane compartment that links the endoplasmic reticulum to mitochondria. During RNA virus infection, RIG-I is recruited to the MAM to bind MAVS. Dynamic MAM tethering to mitochondria and peroxisomes then coordinates MAVS localization to form a signaling synapse between membranes. Importantly, the hepatitis C virus NS3/4A protease, which cleaves MAVS to support persistent infection, targets this synapse for MAVS proteolysis from the MAM, but not from mitochondria, to ablate RIG-I signaling of immune defenses [41].

Mitochondria research is one of the most active areas in medicine. Researchers in the field of not only cancer but also infectious diseases have paid great attention to mitochondria proteomics owing to its potential of providing a better understanding of the protein interactions that regulate diseases phenotype.

References

1. Saito I, Miyamura T, Ohbayashi A, Harada H, Katayama T, Kikuchi S, Watanabe Y, Koi S, Onji M, Ohta Y, Choo QL, Houghton M, Kuo G. Hepatitis C virus infection is associated with the development of hepatocellular carcinoma. *Proc Natl Acad Sci U S A* 1990, 87: 6547–6549.

2. Ploss A, Evans MJ, Gaysinskaya VA, Panis M, You H, de Jong YP, Rice CM. Human occludin is a hepatitis C virus entry factor required for infection of mouse cells. *Nature* 2009, 457: 882–886.

3. Jopling CL, Yi M, Lancaster AM, Lemon SM, Sarnow P. Modulation of hepatitis C virus RNA abundance by a liver-specific MicroRNA. *Science* 2005, 309: 1577–1581.
4. Jopling CL, Schutz S, Sarnow P. Position-dependent function for a tandem microRNA miR-122-binding site located in the hepatitis C virus RNA genome. *Cell Host Microbe* 2008, 4: 77–85.
5. Su AI, Pezacki JP, Wodicka L, Brideau AD, Supekova L, Thimme R, Wieland S, Bukh J, Purcell RH, Schultz PG, Chisari FV. Genomic analysis of the host response to hepatitis C virus infection. *Proc Natl Acad Sci U S A* 2002, 99: 15669–15674.
6. Wakita T, Pietschmann T, Kato T, Date T, Miyamoto M, Zhao Z, Murthy K, Habermann A, Kräusslich HG, Mizokami M, Bartenschlager R, Liang TJ. Production of infectious hepatitis c virus in tissue culture from a cloned viral genome. *Nat Med* 2005; 11: 791–796.
7. Moradpour D, Penin F, Rice CM. Replication of hepatitis C virus. *Nat Rev Microbiol* 2007; 5: 453–463.
8. Boulant S, Montserret R, Hope RG, Ratinier M, Targett-Adams P, Lavergne JP, Penin F, McLauchlan J. Structural determinants that target the hepatitis C virus core protein to lipid droplets. *J Biol Chem* 2006; 281: 22236–22247.
9. Suzuki R, Sakamoto S, Tsutsumi T, Rikimaru A, Tanaka K, Shimoike T, Moriishi K, Iwasaki T, Mizumoto K, Matsuura Y, Miyamura T, Suzuki T. Molecular determinants for subcellular localization of hepatitis C virus core protein. *J Virol* 2005; 79: 1271–1281.
10. Shirakura M, Murakami K, Ichimura T, *et al.* E6AP ubiquitin ligase mediates ubiquitylation and degradation of hepatitis C virus core protein. *J Virol* 2007; 81(10): 74–85.
11. Miyamoto H, Moriishi K, Moriya K, Murata S, Tanaka K, Suzuki T, Miyamura T, Koike K, Matsuura Y. Hepatitis C virus core protein induces insulin resistance through a PA28γ-dependent pathway. *J Virol* 2007; 81: 1727–1735.
12. Moriishi K, Okabayashi T, Nakai K, Moriya K, Koike K, Murata S, Chiba T, Tanaka K, Suzuki R, Suzuki T, Miyamura T, Matsuura Y. Proteasome activator PA28gamma-dependent nuclear retention and degradation of hepatitis C virus core protein. *J Virol* 2003; 77: 10237–10249.
13. Li X, Lonard DM, Jung SY, Malovannaya A, Feng Q, Qin J, Tsai SY, Tsai MJ, O'Malley BW. The SRC-3/AIB1 coactivator is degraded in a ubiquitin- and ATP-independent manner by the REGgamma proteasome. *Cell* 2006; 124: 381–392.

14. Hayashi J, Aoki H, Kajino K, Moriyama M, Arakawa Y, Hino O. Hepatitis C virus core protein activates the MAPK/ERK cascade synergistically with tumor promoter TPA, but not with epidermal growth factor or transforming growth factor alpha. *Hepatology* 2000 ; 32(5): 958–961.

15. Yoshida T, Hanada T, Tokuhisa T, Kosai K, Sata M, Kohara M, Yoshimura A. Activation of STAT3 by the hepatitis C virus core protein leads to cellular transformation. *J Exp Med* 2002 September 2; 196(5): 641–653.

16. Ray RB, Martin LM, Meyer K, Ray R. Hepatitis C Virus core protein cooperates with ras and transforms primary rat embryo fibroblasts to tumorigenic phenotype. *J Virol* 1996; 4438–4443.

17. Moriya K, Yotsuyanagi H, Shintani Y, Fujie H, Ishibashi K, Matsuura Y, Miyamura T, Koike K. Hepatitis C virus core protein induces hepatic steatosis in transgenic mice. *J Gen Virol* 1997; 78: 1527–1531.

18. Moriya K, Fujie H, Shintani Y, Yotsuyanagi H, Tsutsumi T, Ishibashi K, Matsuura Y, Kimura S, Miyamura T, Koike K. Hepatitis C virus core protein induces hepatocellular carcinoma in transgenic mice. *Nat Med* 1998; 4: 1065–1068.

19. Koike K, Moriya K, Ishibashi K, Matsuura Y, Suzuki T, Saito I, Iino S, Kurokawa K, Miyamura T. Expression of hepatitis C virus envelope proteins in transgenic mice. *J Gen Virol* 1995; 76: 3031–3038.

20. Koike K, Moriya K, Ishibashi K, Yotsuyanagi H, Shintani Y, Fujie H, Kurokawa K, Matsuura Y, Miyamura T. Sialadenitis resembling Sjögren's syndrome in mice transgenic for hepatitis C virus envelope genes. *Proc Natl Acad Sci U S A* 1997; 94: 233–236.

21. Bach N, Thung SN, Schaffner F. The histological features of chronic hepatitis C and autoimmune chronic hepatitis: a comparative analysis. *Hepatology* 1992; 15: 572–577.

22. Powell EE, Jonsson JR, Clouston AD. Steatosis: co-factor in other liver diseases. *Hepatology* 2005; 42: 5–13.

23. Moriya K, Todoroki T, Tsutsumi T, Fujie H, Shintani Y, Miyoshi H, Ishibashi K, Takayama T, Makuuchi M, Watanabe K, Miyamura T, Kimura S, Koike K. Increase in the concentration of carbon 18 monounsaturated fatty acids in the liver with hepatitis C: analysis in transgenic mice and humans. *Biophys Biochem Res Commun* 2001; 281: 1207– 1212.

24. Shintani Y, Fujie H, Miyoshi H, Tsutsumi T, Tsukamoto K, Kimura S, Moriya K, Koike K. Hepatitis C virus and diabetes: direct involvement of the virus in the development of insulin resistance. *Gastroenterology* 2004; 126: 840–848.

25. Perlemuter G, Sabile A, Letteron P, Vona G, Topilco A, Chrétien Y, Koike K, Pessayre D, Chapman J, Barba G, Bréchot C. :Hepatitis C virus core protein inhibits microsomal triglyceride transfer protein activity and very low density lipoprotein secretion: a model of viral-related steatosis. *FASEB J* 2002; 16: 185–194.

26. Moriishi K, Mochizuki R, Moriya K, Miyamoto H, Mori Y, Abe T, Murata S, Tanaka K, Miyamura T, Suzuki T, Koike K, Matsuura Y. Critical role of PA28g in hepatitis C virus-associated steatogenesis and hepatocarcinogenesis. *Proc Natl Acad Sci U S A* 2007; 104: 1661–1666.

27. Kim KH, Hong SP, Kim K, Park MJ, Kim KJ, Cheong J. HCV core protein induces hepatic lipid accumulation by activating SREBP1 and PPARgamma. *Biochem Biophys Res Commun* 2007; 355: 883–888.

28. Waris G, Felmlee DJ, Negro F, Siddiqui A. Hepatitis C virus induces proteolytic cleavage of sterol regulatory element binding proteins and stimulates their phosphorylation via oxidative stress. *J Virol* 2007; 81: 8122–8130.

29. Tsutsumi T, Suzuki T, Shimoike T, Suzuki R, Moriya K, Shintani Y, Fujie H, Matsuura Y, Koike K, Miyamura T. Interaction of hepatitis C virus core protein with retinoid X receptor-a modulates its transcriptional activity. *Hepatology* 2002; 35: 937–946.

30. Tanaka N, Moriya K, Kiyosawa K, Koike K, Aoyama T. Hepatitis C virus core protein induces spontaneous and persistent activation of peroxisome proliferator-activated receptor alpha in transgenic mice: implications for HCV-associated hepatocarcinogenesis. *Int J Cancer* 2008; 122: 124–131.

31. Tanaka N, Moriya K, Kiyosawa K, Koike K, Gonzalez FJ, Aoyama T. PPAR-a is essential for severe hepatic steatosis and hepatocellular carcinoma induced by HCV core protein. *J Clin Invest* 2008; 118: 683–694.

32. Miyoshi H, Moriya K, Tsutsumi T, Shinzawa S, Fujie H, Shintani Y, Fujinaga H, Goto K, Todoroki T, Suzuki T, Miyamura T, Matsuura Y, Yotsuyanagi H, Koike K. Pathogenesis of lipid metabolism disorder in hepatitis C: polyunsaturated fatty acids counteract lipid alterations induced by the core protein. *J Hepatol* 2011; 54(3): 432–438.

33. Cohen P, Miyazaki M, Socci ND, Hagge-Greenberg A, Liedtke W, Soukas AA, Sharma R, Hudgins LC, Ntambi JM, Friedman JM. Role for Stearoyl-CoA Desaturase-1 in Leptin-Mediated Weight Loss. *Science* 2002; 297: 240–243.

34. Moriya K, Nakagawa K, Santa T, Shintani Y, Fujie H, Miyoshi H, Tsutsumi T, Miyazawa T, Ishibashi K, Horie T, Imai K, Todoroki T, Kimura S, Koike

K. Oxidative stress in the absence of inflammation in a mouse model for hepatitis C virus-associated hepatocellular carcinogenesis. *Cancer Res* 2001; 61: 4365–4370.

35. Moriya K, Miyoshi H, Shinzawa S, Tsutsumi T, Fujie H, Goto K, Shintani Y, Yotsuyanagi H, Koike K. Hepatitis C virus core protein compromises iron-induced activation of antioxidants in mice and HepG2 cells. *J Med Virol* 2010; 82(5): 776–792.

36. Korenaga M, Wang T, Li Y, Showalter LA, Chan T, Sun J, Weinman SA. Hepatitis C virus core protein inhibits mitochondrial electron transport and increases reactive oxygen species (ROS) production. *J Biol Chem* 2005; 280: 37481–37488.

37. Tsutsumi T, Suzuki T, Moriya K, Yotsuyanagi H, Shintani Y, Fujie H, Matsuura Y, Kimura S, Koike K, Miyamura T. Intrahepatic cytokine expression and AP-1 activation in mice transgenic for hepatitis C virus core protein. *Virology* 2002; 304: 415–424.

38. Tsutsumi T, Suzuki T, Moriya K, Shintani Y, Fujie H, Miyoshi H, Matsuura Y, Koike K, Miyamura T. Hepatitis C virus core protein activates ERK and p38 MAPK in cooperation with ethanol in transgenic mice. *Hepatology* 2003; 38: 820–828.

39. Tsutsumi T, Matsuda M, Aizaki H, Moriya K, Miyoshi H, Fujie H, Shintani Y, Yotsuyanagi H, Miyamura T, Suzuki T, Koike K. Proteomics analysis of mitochondrial proteins reveals overexpression of a mitochondrial protein chaperon, prohibition, in cells expressing hepatitis C virus core protein. *Hepatology* 2009; 50(2): 378–386.

40. Ichibangase T, Moriya K, Koike K, Imai K. A proteomics method revealing disease-related proteins in livers of hepatitis-infected mouse model. *J Prot Res* 2007; 6 (7), 2841–2849.

41. Horner SM, Liu HM, Park HS, Briley J, Gale M Jr. Mitochondrial-associated endoplasmic reticulum membranes (MAM) form innate immune synapses and are targeted by hepatitis C virus. *Proc Natl Acad Sci USA* 2011; 108(35): 14590–14595.

Chapter 13

Identification of Biomarkers of Infectious Disease Using Surface-Enhanced Laser Desorption/Ionisation Mass Spectrometry

Shea Hamilton and Paul R. Langford
Section of Paediatrics, Imperial College London,
St Mary's Campus, London, W2 1PG, United Kingdom
p.langford@imperial.ac.uk

13.1 Introduction

Infectious diseases are responsible for a substantial burden of morbidity and mortality worldwide. They are the world's biggest killer of children and young adults with the citizens of developing countries being particularly vulnerable. As well as the high death toll, infectious diseases are disabling. For example, repeated bouts of disease keep children away from school and prevent adults from working, deformities lead to handicap and stigma and chronic infectious disease can lead to other serious debilitating conditions such as cancer, e.g. hepatitis C virus (HCV) and hepatocarcinoma. Disablement is one of the major reasons for the underdevelopment of many countries particularly those in sub-Saharan Africa [2]. Many of the infectious diseases throughout the world are preventable.

Quantitative Proteome Analysis: Methods and Applications
Edited by Kazuhiro Imai and Sam F. Y. Li

ISBN 978-981-4316-51-4 (Hardcover), 978-981-4316-98-9 (eBook)
www.panstanford.com

However, new diseases and threats are continually emerging. Acquired immune deficiency syndrome (AIDS), severe acute respiratory syndrome (SARS) and influenza virus H1N1 are prime examples of infectious diseases that have had major impacts on the world in the last 20 years. Multi-drug resistance in many infectious agents such as tuberculosis and methicillin resistant *Staphylococcus aureus* (MRSA) is a real threat. Global warming is predicted to result in new diseases or the emergence of old scourges [38]. There is a real need for new preventive measures (e.g. vaccines), therapeutics (new classes of antibiotics) and diagnostics. In the latter category, the infectious disease has typically been diagnosed by the presence of the aetiological agent in clinical specimens directly by either microscopy, after culture or the presence of nucleic acids (DNA/RNA), or antibodies or immune reactions that indicate a particular infection. Depending on the disease and infectious agent, there have been varying degrees of success. CD4 count and viral load is a good indicator of AIDS infection. However, in the case of paediatric tuberculosis diagnosis is very difficult primarily due to the low bacterial load of the causative agent *Mycobacterium tuberculosis*. With the availability of new technologies, there has been an increasing focus on identifying host biomarkers specific to a particular infection or a syndrome that is caused by a number of possible infectious agents. This is a natural extension to the extensive use of acute phase reactants such as C-reactive protein (CRP) that are indicative of infection but lack specificity. In addition, biomarkers are being sought that can indicate effectiveness of therapeutic regimens. This review will focus on a high throughput method to identify biomarkers for specific infectious diseases, syndromes caused by different microorganisms and/or monitoring therapy that is called surface-enhanced laser desorption/ionisation mass spectrometry (SELDI). It will focus on the technology, study design, sampling considerations and the advantages and disadvantages of its use. Emphasis will be given to papers published from January 2008 onwards, as the literature prior to this has been extensively reviewed [29, 34], and to examples of its use applied to infectious diseases that are particularly informative. Discussion has been restricted to clinical-based studies involving SELDI analysis of fluids or specific cell types from infected humans or animals. Such studies include those where there may have been sub-culture of cells from infected and/or controls. Only papers where samples are obtained from infected patients and controls, whether they be analysed directly or from cultured cells, will be considered. For the convenience of readers, we have formulated a summary table of all of the studies discussed in this review in Section 13.9.

13.2 SELDI Technology

SELDI was developed in the 1990s, Ciphergen being the company that commercialized the technology. Following a takeover, Bio-Rad now sells the ProteinChip Reader (mass spectrometer), associated software and consumables. SELDI is a variation of matrix-assisted laser desorption ionisation (MALDI) mass spectrometry. The main difference is the different surfaces that samples are applied to. In MALDI, the surfaces, e.g. stainless steel used are chosen to bind all of the proteins in that sample. This contrasts with SELDI where a range of surfaces (different ProteinChips) are used. The most common for biomarker discovery in the infectious disease arena are hydrophobic reverse phase (H4 and H50), strong anionic exchange (SAX2, Q10), weak cationic exchange (WCX2, CM10) and immobilized metal affinity chromatography (IMAC30). The latter ProteinChips are similar to IMAC columns and can be loaded with different ions, e.g. copper, zinc or rubidium. The equivalent general protein-binding ProteinChips to those used in MALDI are NP20, which have a surface comprising silicon dioxide or alternatively the ProteinChip gold array. The latter is not widely used because of expense. In addition, there are the reactive surface arrays. These include PS10, which has preactivated reactive acyl imidazole moieties, PS20 with preactivated epoxy surface groups and the RS100 with preactivated acyl imidazole groups. Through the reactive groupings, it is possible to covalently attach other proteins such as antibodies allowing either quantitative or qualitative analyses of protein-protein interaction. The PG20 ProteinChip comprises PS20 arrays precoupled with recombinant Protein G. Antibodies recognizing known ligands can be added allowing a quantitative measurement of the ligand of interest. There are comparatively few examples of the use of reactive surfaces in the literature. This reflects that the structure and confirmation of the primary antibody and ligand seems crucial. Some combinations work well, others not at all.

After the sample is placed onto the ProteinChip surface, a series of washes in a buffer of defined pH and salt concentration is carried out. Alternatively, an organic solvent may be used in the case of the hydrophobic ProteinChips. Washing with different buffers will enrich for a sub-set of proteins dependent on their isoelectric point and/or hydrophobicity. For a single ProteinChip type, different mass-spectrometric profiles can be generated by buffering at differing pH. Next, the proteins remaining on the ProteinChip surface are co-crystallised with an energy-absorbing matrix. The type of matrix used primarily depends on the mass range of interest but it is normally sinapinic acid (SPA) or cyano-4-hydroxy cinnamic acid (CHCA). The ProteinChip containing the protein sample co-crystallised with energy-absorbing matrix is placed into the SELDI mass spectrometer and the laser applied. Proteins are ionised and the mass/charge (m/z) values are calculated from the time the ions take to travel down the flight tube of the instrument to the detector. The readout is relative intensity (X-axis) versus m/z ratio (Y-axis). However, in many papers m/z is reported as Da. We have chosen to use m/z or Da dependent on the term used by the author of the paper being discussed. The process and representative spectra are depicted in Fig. 13.1. Using robotics, a single instrument can read hundreds of ProteinChip spots in a day.

13.3 Considerations for a Successful SELDI Study

One of the many advantages of SELDI is that a variety of different sample types can be run. These have included serum, plasma, cell lysates, urine, tears, nipple aspirates, saliva, dialysis fluid, sputum and bronchial alveolar lavage fluid. The advantages and disadvantages of all of these are considered in recent reviews [19, 61]. SELDI is particularly useful when only small volumes of bodily fluid (microlitres), e.g. from small babies and large numbers of samples are available. Study design is also crucial and has been discussed in a number of reviews [12, 13, 19, 35, 42].

A classic study design is illustrated in Fig. 13.2. In brief, some of the considerations that should be made when carrying out a study to identify biomarkers to discriminate between patient groups include the following: (1) Standardised protocols to collect the biological samples and appropriate storage methods. There may be protein degradation in biological samples at particular temperatures. For example, the potential biomarker cystatin C degrades if samples are stored at −20°C [13]. The addition of protease inhibitors has been used by some investigators but this may lead to the loss of potential

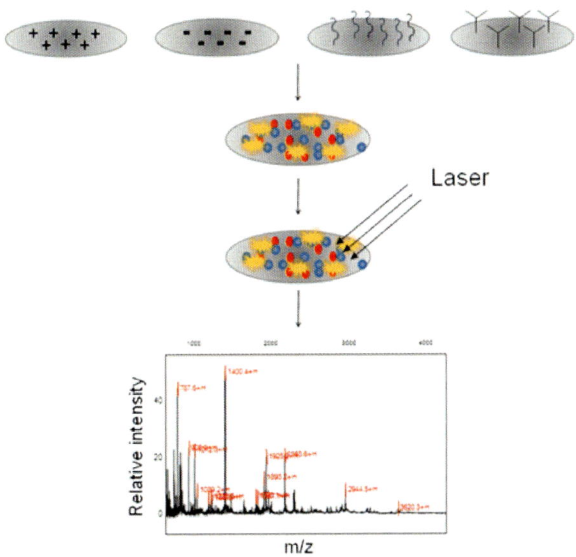

Figure 13.1 The SELDI process. As schematically shown on the top row, different ProteinChips have different surface chemistries. Sample is added to the ProteinChip surface and unbound proteins are removed by washing, leaving a sub-proteome (blue and red dots) of sample remaining on the ProteinChip surface. Energy absorbing matrix (yellow stars) is applied. The ProteinChips are inserted into the SELDI mass spectrometer, the laser applied and a readout of relative intensity versus mass/charge (m/z) ratio is obtained.

biomarkers. (2) Sample preparation and handling. In general, the more the sample is processed prior to analysis the more there is for potential bias through liquid handling errors. For example, there is a greater use of approaches that result in the removal of albumin and/or abundant proteins in serum. The use of robots may be advantageous in this regard. (3) Randomization of samples across ProteinChips and the use of internal controls not only to calibrate the SELDI machine but also of the sample type under investigation. For example, in a study identifying serum biomarkers, pooled serum can be used as an internal control across ProteinChips. (4) The SELDI operators should be blind to the identity of the samples being investigated in the test and validation sets. (5) Many studies have, because of availability of samples, only used small numbers

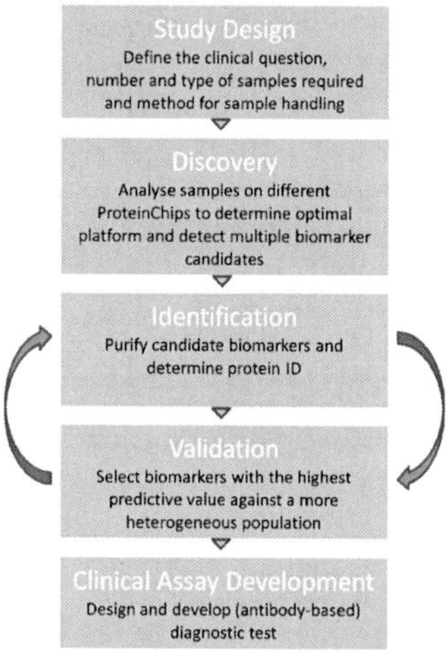

Figure 13.2 An overview of the biomarker discovery process. Arrows indicate that the identification and validation steps commonly occur simultaneously.

of samples. This may be suboptimal. Ideally, different groups (test, control etc.) of at least 30–50 should be used. (6) A typical study design would comprise an initial study on a training set to identify possible biomarkers and subsequent analysis of a test set. Ideally, a separate independent validation cohort is then analysed. However, there are comparatively few papers in which such cohorts are reported. (7) Investigators will also need to consider what their control groups are. The use of healthy controls as a comparative set may lead to the identification of many biomarkers but clinically the use of a cohort, e.g. of patients with diseases that can be mistaken for the one under investigation may be more meaningful. (8) Instrument considerations. Attention should be paid to maintenance, internal calibration and machine settings. (9) Data analysis. There are many methods and software packages available to analyse SELDI data. Readers are referred to recent reviews for further details [18, 19]. Typically, data will be normalized prior to analysis. If a particular biomarker is found to be discriminatory by more than one analytical method, then greater confidence can be assigned to its significance and it will be a good candidate for follow up studies such as molecular identification.

Thus, there are many factors that need to be taken into account when designing a SELDI-based biomarker discovery study. The limitation of potential bias, by taking into account the factors listed above, is crucial to the successful application of the technology.

13.4 Diseases or Syndromes Associated with Viral Infection

13.4.1 Hepatitis B and C Viruses and Hepatocellular Carcinoma

Hepatocellular carcinoma (HCC) is a primary malignancy which is a leading cause of cancer worldwide. The major risk factors are cirrhosis, primarily associated with alcoholism, and hepatitis B virus (HBV) or HCV infection. There have been many proteomic studies searching for biomarkers of HCC because it is difficult to diagnose particularly early stage disease. Additionally, there is a poor prognosis; over 50% of patients may die within a year of definitive diagnosis [47]. The most commonly used biomarker for diagnosis of HCC is serum alpha-fetoprotein (AFP). However, tests that utilise AFP are not sensitive (less than 60%) or specific (less than 90%). Liver biopsy is expensive and potentially deleterious to health. Thus, the ability to identify a highly specific and sensitive biomarker from a readily available body fluid such as serum would be of great benefit. There have been a number of excellent reviews of proteomic studies involving HCC biomarker discovery [21, 34, 46, 65, 68, 71]. With the exception of Zhang *et al.* [74], who investigated liver biopsy material, all of the papers published in 2008 and beyond in this area involve the search for serum biomarkers. All of the papers that conform to our inclusion criteria are detailed below.

Cui *et al.* [17] analysed serum samples obtained from patients with HBV-associated HCC (N = 81), cirrhosis (N = 36) and chronic hepatitis B (N = 43) patients on WCX2 ProteinChips. Between the HCC and liver cirrhosis patients, there were 87 peaks that were significantly different in their expression, as measured by relative intensity. Forty-five peaks had greater than two-fold differences of which 15 were up-regulated and 30 down-regulated. There were only nine peaks (two up- and seven down-regulated) whose intensities differed significantly at the two-fold level when HCC and chronic hepatitis B samples were compared. Finally, when cirrhosis and chronic hepatitis B samples were compared, there were 28 peaks (17 up- and 11 down-regulated) which were significant at the two-fold level. The authors concluded that the five common down-regulated proteins in HCC serum (2870, 3941, 2688, 3165 and 5483 m/z) and the two common up-regulated proteins (3588 and 2017 m/z) in HCC and cirrhosis sera were potentially good biomarkers, but further validation and development work was clearly needed.

He *et al.* [32] compared the sera of those with HBV-related HCC ($N = 81$) and healthy controls ($N = 33$). The training set consisted of 33 HCC and 33 controls and the testing set had 48 HCC and 33 controls. Samples were analysed on WCX2 ProteinChips and decision tree analysis was performed using six proteomic peaks (3157, 4177, 4285, 4301, 7790, and 7984 m/z). The latter peak showed a sensitivity and specificity of 95.92%, and 100%, respectively, and was identified as neutrophil-activating peptide 2. Immunohistochemical staining of a single specimen showed that neutrophil-activating peptide 2 was present in HCC but not adjacent tissues. The authors speculated that neutrophil-activating peptide 2 was a molecule whose up-regulation was liver cancer-specific.

He *et al.* [33]—the first author being different to that in the study described above—analysed the sera of three groups: HBV-related HCC ($N = 67$), HBV infection ($N = 50$), and healthy controls ($N = 44$). A wide range of ProteinChip type was used: IMAC30, WCX2, SAX2, H50, and NP20. Hierarchical clustering and step-wise ordinal logistic regression analyses were applied and three highly discriminatory peaks at 5890, 11,615, and 11,724 m/z that could distinguish HBV-related HCC from HBV infected individuals were identified. In the test set, the three markers could distinguish the two groups with 100% sensitivity and specificity. Two-dimensional gel electrophoresis and peptide mass fingerprinting of the proteins in the 12 kDa region identified a peak of 11,615 Da as being serum amyloid A. Identification of the latter was confirmed by immunodepletion experiments. Serum amyloid A is an acute phase reactant and widely used as a marker of general infection. Thus, it is unlikely to be useful as a biomarker for HCC. However, it remains possible that the two other markers may be of practical use for HCC surveillance and prediction.

In contrast to other authors, Zhang *et al.* [74] analysed liver biopsy material. Samples came from 26 HCC (HBV-related) and 18 liver cirrhosis (non-HBV related) patients and were analysed on CM10 ProteinChips. Sixteen peaks (seven up-regulated, nine down-regulated) could discriminate HCC and cirrhotic (non-HBV related) patients. Discrimination between moderately or well differentiated and poorly differentiated HCC (as determined by histology) was not possible. There was a correlation between histological staining of AFP and biomarker intensity in some cases. A 4674 Da peak was putatively identified as Bax—a protein which activates caspase 3 resulting in nuclear fragmentation (apoptosis). Identification was made by interrogating the ExPASy database using the parameters m/z +/−0.1% and assuming the isoelectric point was 9 +/− 5. There was a correlation between Bax expression, as determined histochemically, and 4674 Da peak intensity. Whilst there was not definitive identification of the 4674 Da peak as Bax, the results were suggestive. This study illustrates the potential merits of a simple search to identify possible candidates to be taken forward to the next stage of analysis. Considerations would include potential cellular location, prior literature and the biological context under investigation.

Wu *et al.* [72] analysed serum samples from 29 HBV-related HCC versus 30 HBV-related liver cirrhosis patients on IMAC30 ProteinChips loaded with copper. Preliminary experiments established that the use of sodium acetate at pH 4.0 was optimal. Forty five protein/peptide peaks were found that differed between the two patient groups. Discriminant analysis was applied and the most significant differentiating peak was one at 3892 Da. This peak was able to distinguish HCC from liver cirrhosis at a sensitivity of 69% and a specificity of 83.3%. The 3892 Da peak was present in six HCC patients who had a negative serum AFP. Further analysis established that a combination of peaks at 9297 and 29,941 Da and AFP gave a sensitivity, specificity and positive predictive value of 82.8%, 93.3% and 92.3%, respectively, when compared with AFP alone. A peak of 9294 Da was one of four serum biomarkers identified in a SELDI-based study of Schwegler *et al.* [59] to discriminate normal, HCV, HCV-related cirrhosis and HCV-related HCC. It is highly likely that the 9294 Da and 9297 Da peaks are the same, although this remains to be proven formally, suggesting that the cognate peptide/protein is not specific to HBV or HCV-related HCC. The authors concluded that the 3392 Da peak was a good complementary diagnostic marker to positive AFP for HCC and also for the diagnosis of AFP negative HCC. None of the molecular identities of peaks of interest was determined.

A relatively large study was carried out by Wu *et al.* [73]. Eighty-one patients with HBV-related HCC were compared with 80 healthy controls. Patients were randomly assigned to a training set (48 HCC, 47 controls) and a testing set (33 HCC, 33 controls). Samples were analysed on WCX2 ProteinChips at pH 4.0 and decision tree analysis using Biomarker Patterns software was used to identify the most informative peaks which were those at 4096 and 7860 Da. In the test set, a sensitivity of 90.9%, specificity of 87.9% and positive predictive value of 81.8% were found. In a multivariate model that included a cut-off of AFP at 400 ng/mL and the two biomarkers, a sensitivity of 92.7% and specificity of 92.7% were found. The sensitivity and specificity if AFP was used alone was 69% and 83%, respectively. The two informative peaks were identified at the molecular level after their separation by gel filtration and/or reverse phase chromatography, and peptide mass fingerprinting. The 4096 and 7860 Da peaks were human thrombin light chain and human-related oncogene-alpha (GRO-alpha), respectively. Neither protein had been reported as biomarkers in previous HCC-based studies. SELDI-based immunoassays were developed for both proteins with the serum levels in each case being significantly higher in HCC patients than healthy controls ($P < 0.01$). GRO-alpha expression was significantly higher in patients with stage III compared with stage I and II disease ($p = 0.04$). Whilst the two biomarkers clearly have diagnostic potential, particularly when compared with AFP alone, further larger studies in different patient cohorts are warranted. It remains to be determined whether the specificity and/or sensitivity obtained using the two biomarkers (with or without AFP) will be maintained when patients with other liver conditions, rather than healthy controls, are used as the comparative group.

A further study was that of Chen *et al.* [16], who investigated two patient cohorts (each $N = 120$): one with HCC and the other with liver cirrhosis. Patients in both cohorts were positive for hepatitis B antigen. Serum SELDI profiles were obtained on IMAC30 ProteinChips and decision tree analysis carried out using Biomarker Patterns software. Training and test sets comprised 60 randomly chosen samples from each of the patient cohorts. In the training set using five peaks (3324, 3994, 4665, 4795 and 5152 Da), decision tree analysis differentiated HCC from LC with a sensitivity and specificity of 98% and 95%, respectively. In the blind test set, a sensitivity of 83%, a specificity of 92% and an accuracy rate of 87.5% were obtained. Impressively, 87% and 89% of HCC patients in stages I and II of disease, respectively were detected. The diagnostic odds ratio indicated that SELDI-profiling was superior too using an APF cut-off of 20 ng/mL (92.7 vs. 9.1). The use of both SELDI-profiling and AFP enhanced the detection rate of HCC considerably (sensitivity 95%, specificity 98% and diagnostic odds ratio of 931). Ten patients were recruited to obtain serum samples prior to and 3 months after surgery. A decrease in the serum level of peak 3324 Da was observed although this was not statistically significant (p = 0.13) There was, however, an increase in relative intensity of the 4795 Da peak (p = 0.01). Presumptive identifications were made of the 3324 Da and 4795 Da peaks as relaxin B chain and PACA-related peptide, respectively. Microarray analysis of 10 pairs of HCC and adjacent non-tumour tissues suggested that RLN2 (relaxin B) gene expression was up-regulated in HCC compared with controls. However, confirmation that the 3324 Da peak is relaxin B is required.

Since 2007, two studies have focused on HCV-related HCC. The first is that of Zinkin *et al.* [75] who analysed the serum of patients with HCC (N = 41) and those with hepatitis C cirrhosis and no HCC (N = 51). The samples were randomly assigned to training (26 cirrhosis, 20 HCC) and validation (25 cirrhosis, 19 HCC) sets. Analysis was done on three different ProteinChip types: CM10, IMAC30 and H50 and the aggregate number of peaks detected was 1145. An 11-peak protein signature was derived using hierarchical clustering and comprised three proteins from the CM10, four from the IMAC30 and four from the H50 ProteinChips. These 11 peaks could distinguish between HBC-related cirrhosis and HCC with a sensitivity and specificity of 85%. The results were compared with standard methods of diagnosis. Using an AFP cut-off of 20 ng/mL, the sensitivity and specificity were 73% and 71%, respectively. With the *Lens culinaris* agglutinin-reactive AFP test, a sensitivity of 63% and specificity of 94% was obtained. A similar sensitivity (84%) but lower specificity (69%) was found when the samples were analysed by the prothrombin induced by vitamin K absence-II test. SELDI identified seven out of eight patients with the largest HCC tumours in contrast to three for AFP, and one each by the *L. culinaris* agglutinin-reactive AFP and prothrombin induced by vitamin K absence-II tests. One of the peaks (13,391 Da) was identified by mass spectrometry as cystatin C. This was verified by immunodepletion and pull down experiments using anti-cystatin C antibodies.

The second study is that of Kanmura *et al.* [39], who investigated the SELDI profiles of serum obtained from 45 HCV-related HCC (HCV-HCC), 42 HCV-related chronic liver disease without HCC (HCC-CLD) and 21 healthy controls on a CM10 surface. When compared with HCV-CLD patients and healthy controls, five protein peaks (4067, 4470, 7564, 7929, and 8130 Da) were significantly increased in HCV-HCC patients. The 8130 Da peak was the most differentially expressed and was identified as a fragment of complement component C3a by mass spectrometry and confirmed by immunodepletion. Another fragment of complement C3a was identified as a marker for HCV-HCC by Lee *et al.* [43] although the m/z of their peak was 8.9 kDa. Complement C3a is know to have a short half-life and be cleaved into C3desarg (8.9 m/z). In the study of Kanmura *et al.* [39], the 8.9 m/z peak was significantly higher intensity in HCV-HCC patients compared with other groups but lacked the discriminatory power compared with the 8.1 m/z peak. Additionally, the 8.9 m/z peak did not significantly increase with time as HCC progressed in individual patients. Kanmura *et al.* [39] found that for discrimination of HCV-HCC patients a sensitivity and specificity of 78% and 52% was obtained using the C3a fragment (8.1 m/z) alone. When combined with AFP and des-gamma-carboxy prothrombin assays, the sensitivity increased to 98% although the specificity dropped to only 20%. No C3a fragment was found in supernatants obtained from culture *in vitro* of either the human hepatocarcinoma HuH-7 or the hepatoblastoma HepG2 cell lines. In contrast to the studies above, Steel *et al.* [62] found C3a to be down-regulated in HBC-infected HCC patients. In the discussion by Kanmura *et al.* [39], they report that serum from 25 HBV-related HCC patients was additionally analysed. Although no data were shown, the authors stated that "35 protein peaks, including the C3a fragment, were over-expressed in the sera of both HCV-HCC and HBV-HCC patients compared to healthy volunteers". Furthermore, "the C3a fragment (8.1 m/z) was particularly overexpressed in the sera of HCVB-HCC patients and was not significantly different between HBV-HCC patients and HCV-CLD patients without HCC". They speculated that the 8.1 m/z peak is a marker of HCV-HCC but not HBV-HCC infection. Clearly, further work is required to substantiate such claims.

Another study to investigate treatment responses of chronic hepatitis C patients is that of Fujita *et al.* [24]. Patients were given pegylated interferon alpha and ribavirin, a regime considered to be the most effective treatment with genotype 1b. Twenty percent of patients are non-virological responders characterised by viral load being positive throughout 48 hours of therapy. It is difficult to predict which patients will respond and those that will be non-virological responders. To identify host factors associated with treatment responses, SELDI profiles were determined on 32 patients with genotype 1b and a high viral load. Independent variables associated with responders were serum ferritin and three peaks (apolipoprotein A1, haemopexin, and transferrin). Impressively, receiver operating characteristic (ROC) curves for prediction of responders using the apolipoprotein A1/haemopexin and haemopexin/transferrin were 0.964 and 0.936. The data show that pre-treatment tests may be of value in identifying chronic hepatitis C patients that respond or do not respond to pegylated interferon alpha and ribavirin. Additionally, the results suggest that there may be a link between treatment resistance and lipogenesis or iron homeostasis in chronic hepatitis C patients.

13.4.2 Epstein–Barr Virus and Nasopharyngeal Carcinoma

Nasopharyngeal carcinoma (NPC) has a higher incidence in parts of Asia compared with Europe and North America. Because of the association between Epstein–Barr virus (EBV) infection and NPC, serological tests detecting antibodies to viral antigens are used for screening high-risk populations. Such antibodies include the EBV viral capsid antigen and nuclear antigen (EBNA1). However, these tests suffer from a lack of specificity and sensitivity. As with many cancers the earlier the disease is detected the better the prognosis. Huang *et al.* [36] analysed the serum profiles of SELDI spectra from a test set of 24 patients with NPC and 24 non-cancer controls and, subsequently, a blind validation set of a further 20 patients with NPC and 12 non-cancer patients. Four ProteinChip surfaces were used CM10, Q10, H50 and IMAC30 loaded with copper. Three peaks (3160, 5188, 13,739 Da) were highly discriminatory. The 13,739 m/z peak was highest in samples from NPC patients whilst the 3160 and 5188 m/z peaks were highest in controls. A decision tree calculated in Biomarker Patterns software using these three peaks resulted in a sensitivity of 95% and specificity of 83.3% for the validation set of samples. None of the peaks was identified at the molecular level. This preliminary study is very promising and further experiments aimed at identifying the biomarkers with a view to formulating a simple clinical test are warranted. From an infectious disease perspective, it will be interesting to determine whether the peptides/proteins of interest are primary (directly related to EBV infection) or secondary biomarkers.

13.4.3 Chronic Fatigue Syndrome

Chronic fatigue syndrome (CFS), also known as post-viral fatigue syndrome (if the onset follows a flu-like illness) or myalgic encephalomyelitis is a debilitating disorder characterised by persistent fatigue not caused by exertion and not relieved by rest. At present, there is no specific diagnostic laboratory test or biomarker for CFS. Eight-eight genes were differentially expressed in the blood of patients with CFS compared with controls and cluster analysis identified seven subtypes with distinct clinical phenotypes and associated disease severity [41]. Amongst those 88 genes were those associated with haematological and immunological function, cancer, cell death and viral infection. It has long been thought that the disease has an infectious disease origin, either as the causative agent or as one of the triggers for subtype(s) of the syndrome. A link between CFS and retrovirus xenotropic murine leukaemia virus-related virus (XMRV) in the U.S.A. was reported by Lombardi [49]. However, subsequent studies throughout the world failed to find such a link. [22, 31, 66, 67]. More recently, Lo *et al.* [48] reported murine leukaemia virus-like retrovirus *gag* gene sequences similar to polytrophic MLVs and XMRVs present in 87% of patients and 7% of blood donor controls. The authors concluded that their data "clearly support" an association of MLV-like viruses with CFS. There are two studies which have either analysed pre-generated [4] or generated [40] SELDI data on CFS patients. Spectra from the sera of 164 subjects (IMAC30, H50 and CM10 ProteinChips) were compared using Bayesian methods with blood transcriptome data obtained from microarray experiments [4]. Peaks of interest were presumptively identified from m/z ratios using ExPASY's TagIdent although, as the authors admit, this approach is potentially flawed because m/z values may be insufficiently accurate and post-translational modifications may have occurred. Nevertheless, in six cases (AHBHD1, MGC29506, PCMTD1, JRK/PSCA and SNRPA1 and TGFB1) there was a correlation between protein and gene expression. The study illustrates the difficulties in correlating SELDI and microarray data without the identity of m/z peaks being known. Typically, in eukaryotic systems, genes that are differentially expressed in a particular disease state may be detected as differentially expressed at the protein level in only 30–70% of cases [28]. In the study of Kerr *et al.* [40], which was not considered in our previous review, serum samples from 30 cases of CFS and 30 normal blood donors (age and sex-matched) were examined on CM10 and Q10 ProteinChips. A potential biomarker of 17,899 Da which was decreased in expression in CFS patients (p = 0.005) was found. This pilot study is promising and a larger study may be valuable. However, whether a single protein, or more likely a cocktail of proteins, would be able to distinguish the 7 subtypes of CFS is a leading question.

13.4.4 HIV-1 Associated Dementia (HAD)

Neurocognitive disorder is a common feature in those infected with HIV. Features might include cognitive impairment such as poor concentration and memory loss, motor symptoms such as clumsiness and poor balance and behavioural changes including apathy and reduced emotions. The symptoms arise after immune activation of brain macrophages and microglia in response to infection. HIV is thought to enter the brain within infected monocyte/macrophages early after infection. The most serious form is HIV-associated dementia (HAD). Highly active antiretroviral therapy can slow disease progression and delay the onset of neurocognitive disorders. However, the prevalence is increasing as more people survive the infection. Unfortunately, there are no biomarkers that predict individuals who will develop HAD. Wiederin *et al.* [69] obtained spectra from serum samples from HIV-1 infected patients with HAD (N = 14) and without (N = 7). All samples were immunodepleted with the Beckman Coulter IgY-12 High Capacity Proteome Partitioning kit, which removes the 12 most abundant serum proteins, before being analysed on WCX2 ProteinChips. Four peaks were significantly differentially expressed between the two groups. Two peaks (4493 and 25,872 m/z) were statistically significantly increased and two decreased (6633 and 25,872 m/z) in patients with HAD. None of the peaks was identified at the molecular level. Ratto-Kim *et al.* [55] analysed the supernatants derived from monocyte/macrophages obtained from a predominantly Thai cohort, and also some individuals from Hawaii, of volunteers with HIV-1 (N = 15) or with HAD (N = 15) and matched HIV-1 negative controls (N = 30). Samples were pre-separated by high-pressure liquid chromatography and fractions analysed on gold ProteinChips. All samples from HIV-1-positive patients were obtained prior to the initiation of antiretroviral therapy. However, no peaks were found that separated HIV-positive patients with and without HAD.

13.4.5 Human Papillomavirus and Oral Squamous Cell Carcinomas

Human papillomavirus (HPV), especially type 16, is associated with a subset of oral squamous cell carcinomas (OSCC). To gain an insight into the pathophysiological events in HPV-positive versus HPV-negative OSCC, a SELDI study was undertaken by Melle *et al.* [52]. Samples comprised formalin-fixed and paraffin-embedded tumours from 17 HPV-associated and 7 HPV non-associated patients. Laser microdissection and a pressure catapulting microscope were used to collect 3000–5000 cells for SELDI analysis on Q10 and CM10 ProteinChips. Eighteen peaks in total differentiated HPV-negative and HPV-positive OSSC. Two of the peaks at 11,434 and 15,339 Da were identified, through the use of two-dimensional gel electrophoresis and peptide mass fingerprinting, as thioredoxin and epidermal-fatty acid binding protein, respectively. Immunodepletion experiments confirmed the identity of both peaks. It was unclear why the two proteins were specifically associated with HPV-associated as opposed to HPV non-associated OSSC. The authors speculated that thioredoxin and epidermal-fatty acid binding protein would be good biomarkers for HPV-associated OSSC.

13.4.6 Porcine Reproductive and Respiratory Disease Syndrome

The porcine reproductive and respiratory disease syndrome (PRSS) virus is responsible for substantial economic losses in the worldwide pig industry through reduced daily weight gain and increased use of medication required for control of secondary pathogens. A combination of antigenic heterogeneity and immune evasion strategies of PRRSV has meant that conventional vaccine strategies have failed to eradicate the virus. PRSS virus targets alveolar macrophages in the lungs and the infection can result in respiratory distress and influenza-like symptoms. The virus can be transferred by secretions and also through the air [37]. Genini *et al.* [25] used SELDI to identify serum biomarkers that could differentiate weaning asymptomatic piglets positive for PRRS viraemia from negative controls (PCR tested) and with potential predictive value for the subsequent occurrence of clinical PRRS. Serum from eight controls and eight infected pigs were analysed on WCX2, IMAC30 and H50 ProteinChips. Eighteen highly significant peaks were found that differentiated infected from control animals. However, none could discriminate infected from controls in an independent validation set. Possible confounders contributing to the results obtained in the training and validation sets were differences in protein concentration and the masking of potential early disease biomarkers by more abundant proteins. Although no biomarkers were discovered, excellent SELDI profiles were obtained on all of the ProteinChip surfaces used.

13.5 Diseases Associated with Bacterial Infection

13.5.1 Tuberculosis

Tuberculosis (TB) is a major global health concern, especially in low- and middle-income countries, and accounted for over 9 million new cases in 2007. Diagnosis is particularly problematic for latent TB and early disease. A positive bacterial culture from sputum remains the "gold standard" for diagnosis; however, results are not obtainable for 2 to 6 weeks. Moreover, a positive bacterial culture is even more problematic in infants due to a paucibacilliary load. A rapid, low-cost, point-of-care test is urgently needed to reliably diagnose TB. Lui *et al.* [45] recently carried out a large study comparing 155 sera samples from smear-positive pulmonary tuberculosis (SPPTB) and smear-negative pulmonary tuberculosis (SNPTB) adults in China. Samples were divided into a preliminary training set (31 SPPTB, 22 SNPTB and 42 non-TB controls) and a blinded testing set (20 SPPTB, 14 SNPTB and 26 non-TB controls) and analysed on a CM10 surface. An optimal decision tree of nine peptide/protein peaks showed a

93.5%, 95.5% and 88.1% accuracy in classifying SPPTB, SNPTB and non-TB controls, respectively. Two peaks (4821 and 4792 m/z) for distinguishing between SNPTB and non-TB controls showed a sensitivity of 86.3% and a specificity of 85.7%. Identification of the 4.8 kDa protein by tandem mass spectrometry revealed a previously unpublished 43 amino acid protein (BOV386_HUMAN) that will require more characterisation to assess its role in the diagnosis of smear-negative cases.

13.5.2 Sepsis

Sepsis is defined as a systemic infection caused by bacterial, viral or fungal infection. Sepsis is the leading cause of death in critically ill patients where severe cases can lead to septic shock and organ failure. Diagnosis of sepsis is complicated and typically based on levels of CRP or procalcitonin (PCT), despite their modest diagnostic specificity. Bacterial sepsis is also the leading cause of death following liver transplantation. In order to search for more reliable biomarkers of sepsis, Paugam-Burtz *et al.* [54] determined the plasma protein profiles of patients with postoperative sepsis (N = 62) compared with non-septic patients (N = 64) on CM10 ProteinChips. The training set, comprising 61 samples, showed that a mean of 317 peaks could be detected on the CM10 surface and a total of 29 were significantly differentially expressed. Sixty-five additional patients (31 with and 34 without sepsis) were used for the test set. The expression of five protein peaks (4152, 4627, 5744, 5812 and 5912 m/z) were common to both data sets and could estimate the likelihood of disease in individual patients better than measuring CRP or PCT plasma levels.

13.5.3 Respiratory Diseases

Chronic obstructive pulmonary disease (COPD) is responsible for substantial morbidity and mortality worldwide. The disease is characterised by progressive and irreversible decline in lung function resulting in reduced airflow in the lungs. COPD is an umbrella term covering two main diseases: chronic bronchitis and emphysema. The most common cause of COPD is smoking, although it is also associated with occupational exposure to pollutants, cold weather and irritants. Disease progression is characterised by intermittent worsening of symptoms called recurrent acute exacerbations, which are typically associated with infections of either viral or bacterial origin, in particular non-typable *Haemophilus influenzae*. A pilot study to screen for candidate biomarkers of acute exacerbations of COPD was reported by Bozinovski *et al.* [6]. Using serum collected from four patients with severe COPD, they found increased expression of an 11.6 kDa peak. Based on the peak mass, they predicted the identity as serum amyloid A, an acute phase inflammatory mediator. This was confirmed by ELISA in a larger cohort (N = 99) of individuals with stable versus acute COPD and longitudinally through convalescence.

Another common respiratory disease is cystic fibrosis (CF), a hereditary disease caused by a mutation(s) in a gene encoding the cystic fibrosis trans-membrane conductance regulator, which is a chloride ion channel. Hallmarks of CF disease include recurrent and persistent bacterial infection. Early infections may be caused by a variety of bacterial pathogens such as *Staphylococcus aureus*, *H. influenzae* and non-mucoid *Pseudomonas aeruginosa*. However, mucoid *P. aeruginosa* infections predominate amongst older patients and these are associated with a substantial decline in lung function. Since 2008, four SELDI studies have been conducted to define specific biomarker signatures that can distinguish between these important respiratory diseases. In CF, airway inflammation is predominantly associated with neutrophils. To further define the immune response, MacGregor *et al.* [50] compared the SELDI proteomic profiles of bronchial lavage fluid from children with CF or other respiratory disease (i.e. lower respiratory tract infection, chronic cough, primary ciliary dyskinesia and croup). Twelve different binding surfaces and wash conditions were used, and 202 proteins/peptides were differentially expressed in the CF samples. The most abundant biomarker was identified as S100A8 (calgranulin A), which was previously reported in induced sputum of CF patients [51]. The most discriminatory biomarker was 5.1 kDa, although this was not identified at the molecular level. There was no correlation between the 5.1 kDa peak and the neutrophil count in BALF, although there was a negative correlation with calgranulin A. Three markers at 6.2, 10.1 (S100A12), and 21.0 kDa, were positively correlated with neutrophil counts.

Some SELDI studies have directly compared patients with COPD and CF and other respiratory diseases. Gray and colleagues [27] compared the proteomic profiles of induced sputum from both COPD (N = 24) and CF (N = 28) patients, 18/28 of the latter who were chronically infected with *P. aeruginosa*. Additional groups comprised healthy controls (N = 20) and patients suffering from bronchiectasis (N = 19) or asthma (N = 24). Longitudinal samples were taken from 12 CF patients during an infective exacerbation, including at onset and after the completion of intravenous antibiotic therapy. Samples were analysed on CM10, Q10 and IMAC ProteinChips to maximise the discovery of new biomarkers. There were a substantial number of differential peaks associated with all disease conditions compared with healthy controls and more similarity was observed between SELDI profiles between the two obstructive diseases and between the two suppurative diseases. Over the three surfaces tested, 58 peaks could differentiate between the groups at $P < 0.01$ and 8 peaks at $P < 0.001$. A number of these peaks were identified, including calgranulin A-C, Clara cell secretory protein, proline rich salivary peptide, lysozyme C, cystatin S and haemoglobin α. CF and bronchiectasis were associated with high levels of calgranulins, including a truncated form of calgranulin A, that was confirmed by western blotting. No correlation between the amount of calgranulin A and neutrophil number was observed. Clara cell secretory protein was lower in abundance in all disease groups and this was confirmed by western blotting. A second study is that of Gomes-Alves *et al.* [26] who investigated the SELDI profiles of both serum and nasal epithelial cells in patients with CF, COPD, asthma and healthy controls. A total of 125 sera were collected (26 CF, 29 asthma, 16 COPD and 54 controls) and run on CM10 ProteinChips. Nasal epithelial cells were collected from 14 CF, 20 asthma and 15 controls and run on both CM10 and Q10 surfaces. Profiling analysis of the serum and nasal epithelial cells revealed over 160 differentiating protein peaks that were statically significant ($P < 0.05$). When logistic regression was applied they observed that combinations of two or more of these peaks significantly improved the predictive value of differentiating these diseases with common clinical features. Protein purification yielded a positive identification of the 15,887 m/z peak as haemoglobin subunit-beta. Both of these studies demonstrate that SELDI shows promise in identifying novel biomarkers in respiratory disease that may lead to advances in our understanding in pathophysiology and could prove useful for monitoring disease progression and treatment.

13.5.4 Intra-Amniotic Infection

Preterm labour, where neonates are born before 37 weeks of pregnancy, is a major cause of morbidity and mortality worldwide. In the Western world, it accounts for 70% of all neonatal deaths and preterm infants are 40 times more likely to die compared with term neonates [5, 15]. Long-term handicaps include blindness, deafness, developmental delay, chronic lung disease, bowel problems (necrotizing enterocolitis) and a higher risk of cerebral palsy. The aetiology of preterm birth is not completely understood; however, intra-amniotic inflammation plays a major role. Infection and/or sepsis are major causes of intra-amniotic inflammation. Organisms including *Fusobacterium nucleatum*, *Mycoplasma hominis* and *Ureaplasma* species have come under particular spotlight but many others have been implicated. There have been a number of studies, in particular from Buhimschi and colleagues, aimed at identifying those at risk of preterm birth using amniotic fluid. These studies established four biomarkers (defensin-1, defensin-2, calgranulin A and calgranulin C) that could accurately predict intra-amniotic inflammation or infection with high sensitivity and specificity [9]. Moreover, these four proteins were used to formulate a mass-restricted (MR) score, which ranged from 0–4 depending on the presence or absence of these protein peaks, or that showed a 100% correlation with the presence of intra-amniotic inflammation. If a peak is present, it is given a value of 1 and if absent a value of 0. Values of 3 or 4 indicate severe intra-amniotic inflammation. SELDI-based studies in this area of research have proved particularly fruitful and best demonstrate where this technology can be used for rapid diagnosis in clinical practice. Studies prior to 2008 have been the subject of previous reviews [29, 34].

A relationship between MR score and histologic chorioamnionitis was established in a study investigating 158 consecutive women with singleton pregnancies admitted with preterm labour or premature rupture of membranes [10]. There was a significant correlation between the MR score, severity of histologic chorioamnionitis and inflammation of the amnion, choriodecidua and chorionic plate, with African-American women being overrepresented in the severe inflammation group. Calgranulin C was the strongest predictor of severe chorioamnionitis, time to delivery and gestational age and independent of mode of delivery or exposure to antibiotics or steroids. Calgranulin C (S100A12) is a calcium-binding protein that is associated with neutrophil function and released during inflammatory responses. Buhimschi and colleagues [8] also did a prospective study to determine the relationship between MR score and foetal inflammatory status at birth in a cohort of 132 patients. The ratio of IL-6 level in cord blood compared with amniotic fluid was used as an indicator of the differential inflammatory response in each compartment. Shorter amniocentesis-to-delivery intervals were found for women who had a MR score of 3–4 and higher IL-6 levels were found in the amniotic fluid of women and in the cord blood of neonates at birth. The IL-6 levels in neonates and the severity of intra-amniotic inflammation could not always be predicted, however. Early onset neonatal sepsis was also associated with increased levels of IL-6 in cord blood. Concentrated amniotic fluid was cultured for anaerobic bacteria in cases where the MR score suggested possible infection. It was not possible, however, to detect bacteria in approximately 30% of cases. The lack of positive bacterial culture, the gold standard for detecting intra-amniotic infection, could be due to the presence of unculturable or difficult to cultivate bacteria. In order to determine the rate of undetected infections in women with intra-amniotic inflammation, Han *et al.* [30] examined the amniotic fluid from preterm patients and controls by bacterial culture and 16S rRNA and related it to the MR score. Amongst the preterm birth samples, bacterial DNA was amplified from all culture-positive samples and, most important, from 17% of the culture negative samples. 16S rRNA analysis also detected further species in the culture-positive samples and there was evidence of intra-amniotic inflammation, as adjudged by the MR score, in the samples which were culture negative but PCR positive.

Buhimschi *et al.* [11] have more recently employed multidimensional proteomics coupled with non-hierarchical bioinformatics algorithms to devise a new ranking system, the Q-profile, which is complementary to the MR score. The Q profile was generated from 286 amniotic fluid samples from women presenting with preterm labour. The proteomic fingerprint consisted of five peaks in the 10 to 12.5 kDa mass range that could identify a novel pathway related to preterm birth in the absence of intra-amniotic inflammation or bleeding. The results suggest that in some patients there is a distinct intra-amniotic inflammation-independent pathway that results in preterm birth. The study illustrates the powerful combination of proteomics and pathway analysis. In a retrospective cross-sectional SELDI study, Romero *et al.* [57] investigated patients with spontaneous preterm labour who delivered at term and those that delivered preterm with intra-amniotic inflammation. They also developed a novel computational method which involved SELDI spectra obtained from two ProteinChip surfaces (CM10 and H50), using two energy-absorbing matrices and two laser intensities. Thirty-nine peaks could discriminate between patients with intra-amniotic inflammation/infection versus those without with a sensitivity and sensitivity of 91.7% and 91.5%, respectively. Some of the m/z values corresponded to those found in previous studies and used in the MR score.

13.5.5 Bacterial Peritonitis

In patients with renal failure, osmotic correction of blood solute can be achieved across the internal lining of the abdomen in a process called continuous ambulatory peritoneal dialysis (CAPD). Although the ability of CAPD to successfully treat end-stage renal disease is well established, infection at the catheter site remains a serious cause of morbidity. In order to identify biomarkers of infection, Lin *et al.* [44] compared peritoneal fluid of patients undergoing CAPD. Several bacteria, including *Staphylococcus epidermidis, S. aureus* or *Klebsiella pneumoniae*, were isolated from the peritoneal fluid of infected patients. A combination of two-dimensional polyacrylamide gel electrophoresis and SELDI found an 11,117 m/z peak that could differentiate infected patients from controls with an accuracy of 95%. The peak was identified as a peptide fragment derived from β-2-microglobulin, (full length 11,731 m/z), and confirmed by SELDI-based immunodepletion experiments. β-2-microglobulin is an established peritoneal fluid biomarker for infection.

13.6 Parasitic Infections

13.6.1 *Fasciola Hepatica*

The trematode *Fasciola hepatica* (common liver fluke) can cause fasciolosis in many herbivorous mammals, including sheep and cattle, and humans. It is found mainly in Europe, America and Oceania and is an economically important pathogen in livestock. Rioux *et al.* [56] evaluated SELDI to identify serum biomarkers associated with *F. hepatica* infection of Corriedale sheep. Sera were collected at weekly intervals from eight experimentally infected sheep and pooled, fractionated and analysed on CM10 ProteinChips. For the test phase, sera from individual sheep (3 and 9 weeks post infection) were used. Six biomarkers were up-regulated at weeks 3 and 9 post infection, 16 were up-regulated only at week 9 post infection and four biomarkers were down-regulated at week 9 post infection. Transferrin and apolipoprotein A-IV were identified as two of the biomarkers that were up-regulated at 9 weeks post-infection. This study has been one of the few that have analysed pooled serum in the initial test phase and individual sera in a validation phase.

13.6.2 Cestode Infection of the Brine Shrimp

The brine shrimp *Artemia parthenogenetica* is an economically important organism and an intermediate host for many species of cyclophyllidean (cestode) tapeworms which can infect several water birds. Cestode eggs are released by tapeworms into water in the faeces of their avian host and consumed by the brine shrimp. The newly hatched oncosphere penetrates the haemocoel and develops into a cysticeroid which is associated with changes of behaviour and colour and makes the brine shrimp more susceptible to being eaten by foraging birds, as they are more likely to be found at the water surface. Adult tapeworms grow in the small intestine of the birds and the life cycle starts again. In order to understand these behavioural changes, Sánchez *et al.* [58] infected brine shrimps with *Flamingolepsis liguloides*, *Confluaria podicipina* and *Anomotaenia tringae* which infect flamingos, grebes and a variety of shore birds, respectively. The five categories investigated were (1) uninfected brine shrimp from the bottom; (2) uninfected brine shrimp from the bottom placed at the water surface (controls); (3) brine shrimp infected by *F. liguloides*; (4) brine shrimp infected by *C. podicipina*; and (5) brine shrimp infected with *A. tringae*. In total, the heads of 56 infected, 52 uninfected from the bottom and 42 uninfected controls brine shrimp heads were analysed on H50, NP20 and IMAC30 ProteinChips. In possibly the most unusual processing procedure in the literature, brine shrimp heads were individually bound to ProteinChips for 24 hr at 4°C prior to SELDI-profiling. A peak at

4515 m/z was down-regulated in brine shrimp brain when infected with *A. tringae* and 3951 m/z when infected with *F. liguloides* and *C. podicipina*. Two peptides (4.1 and 4.2 m/z) were of increased relative intensity in all samples of brine shrimp brains at the surface. The latter suggests that these peptides are associated with the environment rather than infection per se. None of the peptides was identified, although this is the next logical step.

13.6.3 Chaga's Disease

Chaga's disease is caused by the flagellate protozoan *Trypanosoma cruzi* and is transmitted by triatomid blood sucking insect vectors and occasionally through blood transfusion, organ transplantation, ingestion of contaminated food and from mother to foetus. It is most common in the Americas, particularly in poor rural areas of South America. Most infected individuals are asymptomatic and 15–30% will develop cardiac or gastrointestinal complications. Those that do get severe illness after infection will have local swelling and fever. Current testing methods, which are principally immunological in nature, are associated with false positives and there is a need for more accurate tests. Thus, Ndao *et al.* [53] carried out a large study which involved the analysis of sera from asymptomatic Chaga's disease patients (N = 131) and uninfected controls (N = 164) and those with other parasitic diseases (N = 131), including malaria, leishmaniasis and schistosomiasis. Samples were analysed on WCX2 and IMAC30 ProteinChips. Decision tree analysis using Biomarker Patterns software identified six peaks that were consistently found to be included in algorithms that differentiate Chaga's disease patients and controls. For example, a five-node decision tree achieved 98% sensitivity and 100% specificity in separating Chaga's disease patients from the other control groups. The six candidate biomarkers were identified at the molecular level. They were a C-terminal truncation of C3a anaphylatoxin desArg (8.1 m/z), various N-terminal truncations of apolipoprotein A1 (9.3, 10.1, and 13.6 m/z), a C-terminal truncation of apolipoprotein A1 (24.7 m/z) and a C-terminal truncation of fibronectin (28.9 m/z). Because the cost of using SELDI is likely to be prohibitive for routine use, particularly in areas of the world where Chaga's disease is the most common, western blot and ELISAs were evaluated for use as a rapid, simple and cost-effective test. In the case of fibronectin, a 28.9 kDa full length fragment was detected in western blots primarily in Chaga's disease patients compared with those with other parasitic diseases, although only a sensitivity of 78% and specificity of 90% could be achieved. Three biomarkers (6.3, 7.5 and 15.2 m/z) were highly specific to Chaga's disease with cardiac involvement when compared with samples from other parasitic diseases. The authors speculated that these may be associated with early cardiac complications of Chaga's disease and were worthy of specific follow up.

13.6.4 *Taenia Solium*

(*Taenia solium*) is found worldwide but particularly in areas where humans live in close contact with pigs and eat pork that has been undercooked. Eggs or gravid proglottids that are ingested by humans or pigs hatch in the intestine, invade the intestinal wall and can migrate to brain, liver, striated muscle and other tissues where they develop into cystercerci. In humans, infection of the central nervous system can result in neurocysticercosis characterised by headaches, epilepsy, intracranial hypertension and occasionally death. Humans become infected when they eat pork containing the cystercerci. In pigs—an intermediate host—infections are predominantly asymptomatic. There is no diagnostic test that can discriminate between infections with viable cysts (active cysticercosis) and degenerated cysts (inactive cysticercosis). Such a test would be helpful in epidemiological, vaccine and antihelmintic studies. To identify potential biomarkers suitable for use in diagnostic tests, Deckers *et al.* [20] analysed serum samples from naturally infected pigs from Zambia ($N = 27$) and Peru ($N = 59$), experimentally infected pigs ($N = 18$) and pigs infected with *Taenia saginata asiatica* ($N = 6$), *Taenia hydatigena* ($N = 7$), *Trichinella spiralis* ($N = 5$) and *Trypanosoma congolense* ($N = 7$). CM10 and IMAC30 ProteinChips were used. In the experimentally infected animals, 30 discriminating biomarkers were found, 13 being active cysticercosis-specific, nine for inactive cysticercosis and eight common to both groups. Five biomarkers were identified at the molecular level as the C-terminal fragment of the alpha and beta chain of clusterin (3.63 kDa), an N-terminal fragment of vitronectin (12.3 kDa), a C-terminal fragment of Apo-A1 (15 kDa), a N-terminal fragment of lecithin-cholesterol-acyltransferase (22.5 kDa) and an N-terminal fragment of haptoglobin (10.3 kDa). The first four biomarkers were all elevated in pigs with active cysticercosis and the last one with inactive cysticercosis. In the field samples, 10 biomarkers were discriminatory. Only three of the biomarkers (8.6, 28.5 and 55.8 kDa) were found in both experimental and field sample sets, although not consistently. The authors believed that biologically the presence of haptoglobin, clusterin and vitronectin could be linked to cysticercosis infection but that the other two biomarkers identified were probably non-specific consequences of infection.

13.7 Prion Disease

Prion diseases, also known as transmissible spongiform encephalopathies (TSEs), are a group of fatal neurodegenerative diseases that can affect both humans and animals. They include bovine spongiform encephalopathy (BSE) in cows, scrapie in sheep and chronic wasting disease (CWD) in deer. Human forms of prion disease can be genetic (Gerstmann–Sträussler–Scheinker (GSS) syndrome), sporadic (Creutzfeldt–Jakob disease (sCJD)) or infectious (variant CJD). These diseases are characterised by the accumulation of a misfolded protein isoforms $PrP^{s(scrapie)}$ or the prion protein $PrP^{c(cellular)}$ in brain tissue. This conformational change results in increased protein stability and resistance to denaturation, making these proteins very hard to destroy. While definitive disease can only be diagnosed post mortem, specific biomarkers are now used to assign a "probable diagnosis", including 14-3-3-protein or tau protein in the case of CJD. Two recent studies have used SELDI technology to discover new putative biomarkers for TSEs in order to develop a more reliable pre-mortem diagnostic. Steinacker *et al.* [63] compared cerebrospinal fluid from 32 CJD patients, 32 patients having other dementive diseases and 31 non-demented controls on a CM10 surface. Training and test sets comprised 47 randomly chosen samples from each of the patient cohorts. Direct comparison of normalised peak intensities showed that three peaks (6438, 7460, 8606 Da) were significantly differentially expressed ($P < 0.001$) in the training set when comparing all three groups. In the blind test set, only two of these peaks (6438 and 8606 Da) could discriminate between the three groups. Moreover, using the 8.6 kDa peak as a marker of disease resulted in more patients correctly diagnosed when compared with tau or 14-3-3-proteins. This peak was successfully identified through enrichment by reverse phase chromatography followed by gel separation and in-gel trypsin digest. Liquid chromatography tandem mass spectrometry identified this protein as ubiquitin and the increased level of this protein in cerebrospinal fluid of CJD patients was confirmed by western blot.

A further study was that of Barr and colleagues [3] who compared the SELDI profiles of murine brain tissue from groups infected with TSE (scrapie) to non-infected controls. This study examined the protein profiles of infected ($N = 6$) and non-infected ($N = 6$) animals throughout the course of infection (30, 60, 90, 120, 150, 180, 210 and 240 days) and after 260 days ($N = 12$ infected; $N = 12$ controls), clinically assessed as the terminal stage of disease. Samples were analysed on CM10 and Q10 ProteinChips and statistically significant peaks were identified using Biomarker Wizard software. These data were further analysed by cluster analysis and principal component analysis (PCA) using the statistical package *R*. The analysis revealed that total separation between the diseased and the control group was achieved after 90 days of infection. A cluster of four peaks (5235, 6333, 8445 and 16,885 m/z) on the Q10 surface gave a sensitivity and specificity of 100% indicating they would be model biomarkers for distinguishing the healthy from the diseased. Three peptide peaks shown to be significantly differentially expressed on a CM10 surface were identified by fractionating the samples, using 1-D PAGE to isolate the bands of interest and MS/MS applied directly to the ProteinChip surface (see Bio-Rad LUCID system). The protein peaks identified included two that were up-regulated in TSE disease (Hsp/Cpn10 and DBI/ACBP) and one negative biomarker of disease (FKBP12). Differential regulation of these three proteins was also confirmed by immunohistochemistry.

13.8 Hepcidin Measurement

Hepcidin is a protein hormone produced by hepatocytes and is a central regulator of iron homeostasis, negatively regulating duodenal absorption and the iron release from macrophages. It is linked with infection through restriction of iron availability to pathogens, inflammation and has some antimicrobial properties. The active form of hepcidin is a 25 amino acid peptide. N-terminal isoforms (hepcidin-20 and hepcidin-22), truncated at the Nterminus, also exist. However, the role of the isoforms is unclear since the five N-terminal peptides of hepcidin-25 are necessary for binding to ferroportin and biological activity. However, hepcidin-20 has some antimicrobial activity and may play an important role in infection. Hepcidin has proved to be a difficult molecule to assay by conventional techniques because it is small, forms hairpins and is compact with four disulphide bonds. There are few antigenic epitopes and it is highly conserved between species making it difficult to raise antibodies. A number of methods to assay hepcidin are available, e.g. competitive radioimmunoassay, MALDI-TOF and liquid chromatography tandem mass spectrometry (reviewed in 14). Quantitative SELDI analysis has also emerged as a reliable method. Its use has the advantage that the various isoforms can be measured. Analysis is done on IMAC30-copper loaded ProteinChips, with standard curves of hepcidin-25 being calculated and a synthesized hepcidin-24 peptide is used as an internal standard. Examples of where SELDI-based hepcidin assays have been used in the infectious disease arena include studies by Fujita *et al.* [23] and Sugimoto *et al.* [64]. In the former study, serum hepcidin was measured in 73 consecutive chronic hepatitis C (CHC) patients and compared with patients with anaemia of inflammation (N = 10) and age-matched healthy controls (N = 10). Additionally, serum hepcidin was measured in 27 of the CHC patients treated with a 48-week course of pegylated-interferon plus ribavirin therapy. Serum hepcidin-to-ferritin ratios were significantly lower in HCV-positive patients versus control groups. Hepcidin levels returned to normal after successful HCV eradication with the antiviral therapeutic regimen. Sugimoto *et al.* [64] measured serum hepcidin levels and sequential changes during phlebotomy in chronic hepatitis C patients (N = 9) and healthy controls (N = 10). Serum hepcidin were significantly higher in HCV patients than controls and phlebotomy significantly decreased serum hepcidin in HCV patients. One potential biomarker of sepsis is hepcidin, a liver derived peptide that is detectable in both serum and urine and strongly induced by inflammation and sepsis. A recent study by Altamura *et al.* [1] showed that a CM10 surface to accurately and sensitively detect hepcidin in the urine of patients suffering from severe sepsis (N = 8) compared with healthy controls (N = 12). Further work will be required to establish the sensitivity and specificity of hepcidin in the diagnosis of sepsis in the background of other pathologies and inflammatory disorders.

Table 13.1 Summary of the patient/control populations investigated, sample preparation and data analysis methods used in the referenced studies

Disease or aetiological agent	Type and number of samples analysed	Sample preparation	Protein Chip surface(s) used	Data analysis	Addition of clinical data	Validation	Peak ID	Ref
Hepatitis B	HCC = 81 Healthy controls = 80	Serum denatured with CHAPS/urea	WCX2	Decision tree	No	Independent validation set HCC = 33 Healthy controls = 33	No	[73]
Hepatitis B	HCC = 120 LC = 120	Serum denatured with CHAPS/urea	IMAC30	Decision tree	Yes	Independent validation set HCC = 60 LC = 60	No	[16]
Hepatitis B	HCC = 67 HBV = 50 Normal = 44	Serum denatured with CHAPS/urea	IMAC30	Hierarchical clustering and logistic regression	No	Independent validation set HCC = 17 HBV = 8 Normal = 14 Biomarker identified: SAA	Yes	[33]

Disease or aetiological agent	Type and number of samples analysed	Sample preparation	Protein Chip surface(s) used	Data analysis	Addition of clinical data	Validation	Peak ID	Ref
Hepatitis B	HCC = 81 Cirrhosis = 36 LC = 43	Serum denatured with CHAPS/ urea	WCX2	Direct comparison of normalized peak intensities	No	No	No	[17]
Hepatitis B	HCC = 81 Controls = 33	Serum	WCX2	Decision tree	No	Cross validation Independent validation set HCC = 48 Controls = 33 Biomarker identified: neutrophil activating protein-2	Yes	[32]
Hepatitis B	HCC = 26 Cirrhosis = 18	Liver biopsy	CM10	Direct comparison of normalized peak intensities	No	No Biomarker inferred from mass: Bax (expression confirmed by immunohistochemistry)	No	[74]

(*Continued*)

Table 13.1 (*Continued*)

Disease or aetiological agent	Type and number of samples analysed	Sample preparation	Protein Chip surface(s) used	Data analysis	Addition of clinical data	Validation	Peak ID	Ref
Hepatitis B	HCC = 29 LC = 30	Serum denatured with CHAPS/urea	IMAC30	Discriminant analysis	Yes	Cross validation	No	[72]
Hepatitis C	HCC = 41 HCV cirrhosis = 51	Serum	CM10 IMAC30 H50	Hierarchical clustering	Yes	Independent validation set HCC = 19 HCV cirrhosis = 25 Biomarker identified: cystatin C	Yes	[75]
Hepatitis C	HCV-HCC = 45 HCC-CLD = 42 Healthy controls = 21	Serum	CM10	Direct comparison of normalized peak intensities	Yes	Biomarker identified: complement component 3a (C3a) - confirmed by immunodepletion	Yes	[39]

Disease or aetiological agent	Type and number of samples analysed	Sample preparation	Protein Chip surface(s) used	Data analysis	Addition of clinical data	Validation	Peak ID	Ref
Hepatitis C	CHC = 32	Serum denatured with CHAPS/ urea Q-Hyper D anion-exchange resin fractionation	CM10 Q10 IMAC30	Multiple logistic regression	Yes	Biomarkers identified: apolipoprotein A1, apolipoprotein C1, albumin, haemopexin, transferrin, CTAP-III, platelet factor-4	Yes	[24]
EBV	NPC = 24 Healthy controls = 24	Serum denatured with CHAPS/ urea	CM10 Q10 H50 IMAC30	Decision tree	No	Independent validation set: NPC = 20 Healthy controls = 12	No	[36]
HIV-1	HAD = 14 Negative controls = 7	Immunodepleted serum	WCX2	Logistic regression	No	No	No	[69]

(Continued)

Table 13.1 (*Continued*)

Disease or aetiological agent	Type and number of samples analysed	Sample preparation	Protein Chip surface(s) used	Data analysis	Addition of clinical data	Validation	Peak ID	Ref
HIV-1	HIV-infected no dementia = 15 HIV-infected with dementia = 15 Matched HIV-negative controls = 15	Secreted proteins from cultured macrophages / monocytes CHAPS/urea	Gold	Direct comparison of normalized peak intensities	No	No	No	[55]
Chronic fatigue	CFS = 30 Age matched healthy controls = 30	Serum	CM10 Q10	Direct comparison of normalized peak intensities	No	No	No	[40]
HPV	OSCC + HPV = 17 OSCC - HPV = 7	Tissue section	Q10 CM10	Direct comparison of normalized peak intensities	No	No. Biomarkers identified: thioredoxin, epidermal-fatty acid binding protein	Yes	[52]

Disease or aetiological agent	Type and number of samples analysed	Sample preparation	Protein Chip surface(s) used	Data analysis	Addition of clinical data	Validation	Peak ID	Ref
PRRS	Controls = 8 Infected = 8	Serum	WCX IMAC30 H50	Direct comparison of normalized peak intensities	No	Cross validation Controls = 16 Infected = 16	No	[25]
Tuberculosis	Smear + PTB = 51 Smear – PTB = 36 Non-TB controls = 68	Serum denatured with CHAPS/urea	CM10	Decision tree	Yes	Independent validation set: Smear + PTB = 20 Smear – PTB = 14 Non-TB controls = 26 Biomarker identified: Previously unreported protein "BOV386_HUMAN"	Yes	[45]
Sepsis	Infected = 62 Non-infected = 64	Serum/Plasma	CM10	Logistic regression	Yes	Independent validation set: Infected = 31 On-infected = 34	No	[54]

(*Continued*)

Table 13.1 (*Continued*)

Disease or aetiological agent	Type and number of samples analysed	Sample preparation	Protein Chip surface(s) used	Data analysis	Addition of clinical data	Validation	Peak ID	Ref
Sepsis	Infected = 8 Non-infected = 12	Urine	NP20 CM10	Direct comparison of normalized peak intensities	No	Known protein. Validation not required. Hepcidin-25 (2789.4 Da) Truncated Hepcidin-22 (2436 Da) Truncated Hepcidin-20 (2191.7 Da)	Yes	[1]
CF	CF = 39 Other respiratory diseases = 38	BALF	CM10 Q10 IMAC30	Direct comparison of normalized peak intensities	No	No Biomarker identified: S100 A8 (calgranulin A). Correlation of two biomarkers with % bronchoalveolar neutrophil numbers	Yes	[50]
COPD	COPD = 4 Number of controls not reported.	Serum denatured with CHAPS/ urea	H50	Direct comparison of normalized peak intensities	No	No Biomarker predicted mass as serum amyloid a and confirmed by ELISA	Yes	[6]

Disease or aetiological agent	Type and number of samples analysed	Sample preparation	Protein Chip surface(s) used	Data analysis	Addition of clinical data	Validation	Peak ID	Ref
CF + COPD	Serum: COPD = 16 CF = 26 Asthma = 29 Healthy controls = 54 Nasal epithelial cells: CF = 14 Asthma = 20 Healthy controls = 15	Serum denatured with CHAPS/ urea Nasal epithelial cells	CM10 Q10	Logistic regression	No	Serum: COPD = 6 CF = 8 Asthma = 8 Healthy controls = 12 Nasal epithelial cells: CF = 6 Asthma= 6 Healthy controls = 6 Biomarker identified: haemoglobin subunit-beta	Yes	[26]
CF + COPD	Healthy controls = 20 Asthma = 24 COPD = 24 CF = 28 Bronchiectasis = 19 CF patients sampled before and after antibiotic therapy for an infective exacerbation = 12	Induced sputum	CM10 IMAC30 Q10	Direct comparison of normalized peak intensities	No	No Biomarkers identified: calgranulin A-C, Clara cell secretory protein, proline rich salivary peptide, lysozyme C and Cystatin S. There were decreases in the levels of calgranulin A and B (measured by SELDI) and calprotectin (heterodimer of calgranulins A and B) measured by ELISA	Yes	[27]

(*Continued*)

Table 13.1 (*Continued*)

Disease or aetiological agent	Type and number of samples analysed	Sample preparation	Protein Chip surface(s) used	Data analysis	Addition of clinical data	Validation	Peak ID	Ref
IAI Histologic chorioamniotis	PPROM + PTL = 158 Newborns = 125	AF	H4	Direct comparison of normalized peak intensities - mass restricted (MR) score of 4 proteins	No	No	No	[10]
IAI	Singletons with symptoms of PTL No inflammation and no bleeding = 193 Inflammation = 71 Bleeding = 6 Inflammation and bleeding = 15	AF	H4	Direct comparison of normalized peak intensities - mass restricted (MR) score of 4 proteins Q-profile (5 peaks in 10 to 12.5 kDa) range	No	No Description of Q-profile Biomarkers identified: numerous including apolipoprotein A-I IFGBP-1 apolipoprotein A-IV lumican 1 IgM cold agglutinin α-1-antitrypsin	Yes	[11]

Disease or aetiological agent	Type and number of samples analysed	Sample preparation	Protein Chip surface(s) used	Data analysis	Addition of clinical data	Validation	Peak ID	Ref
IAI	Consecutive mothers = 132 MR score 0 = 26 MR score 1–2 = 39 MR score 3–4 = 67 Umbilical cord blood	AF	H4	Direct comparison of normalized peak intensities - mass restricted (MR) score of 4 proteins	No	No Correlation of MR score with clinical parameters	No	[8]
IAI	Term = 59 PTL + IAI = 60	AF	CM10 H50	Binary fingerprinting and pattern discovery	No	Cross-validation Independent validation set Term = 15 PTL + IAI = 16	No	[57]
Peritonitis	Infected = 5 Controls = 11	Peritoneal dialysis fluid	CM10 IMAC30	Direct comparison of normalized peak intensities	No	No Biomarker identified: SAA	Yes	[44]

(Continued)

Table 13.1 (*Continued*)

Disease or aetiological agent	Type and number of samples analysed	Sample preparation	Protein Chip surface(s) used	Data analysis	Addition of clinical data	Validation	Peak ID	Ref
Cysticerosis	Naturally infected = 86 Experimentally infected = 30 (16 controls prior to infection, 14 infected)	Serum denatured with CHAPS/urea Q-Hyper D anion-exchange resin fractionation	CM10 IMAC30	Direct comparison of normalized peak intensities	No	No Biomarkers identified: C-terminal fragment of the alpha and beta chain of clusterin, N-terminal fragment of vitronectin, C-terminal fragment of Apo A-1, N-terminal fragment of LCAT and the N-terminal fragment of haptoglobin	Yes	[20]
Chaga's disease	Chaga's = 131 Uninfected controls = 164 Other parasitic diseases = 131	Serum denatured with CHAPS/urea Q-Hyper D anion-exchange resin fractionation	WCX2 IMAC30	Direct comparison of normalized peak intensities + Decision tree analysis	No	Independent validation set: Chaga's = 79 Uninfected controls = 182 Biomarkers identified: C-terminal C3a desArg; N-terminal apolipoprotein A1; C-terminal apolipoprotein A1; C-terminal fibronectin. Expression of 3 proteins confirmed by ELISA or depletion	Yes	[53]

Disease or aetiological agent	Type and number of samples analysed	Sample preparation	Protein Chip surface(s) used	Data analysis	Addition of clinical data	Validation	Peak ID	Ref
Fasciolosis	Infected = 8 Serial bleeds 0–12 weeks	Serum denatured with CHAPS/urea Q-Hyper D anion-exchange resin fractionation	CM10	Direct comparison of normalized peak intensities	No	No Biomarkers identified: transferrin and apolipoprotein A-IV	Yes	[56]
Cestode	Infected = 56 Uninfected from bottom = 52 Uninfected from surface = 42	Brine shrimp heads	H50 NP20 IMAC30	Direct comparison of normalized peak intensities	No	No	No	[58]
TSE (CJD)	CJD = 32 Other dementive disease = 32 Non-demented controls = 31	CSF	CM10	Direct comparison of normalized peak intensities	No	Independent validation set: CJD = 16 Other dementive disease = 16 Non-demented controls = 16 Biomarker identified: ubiquitin	Yes	[63]

(*Continued*)

Table 13.1 (*Continued*)

Disease or aetiological agent	Type and number of samples analysed	Sample preparation	Protein Chip surface(s) used	Data analysis	Addition of clinical data	Validation	Peak ID	Ref
TSE (scrapie)	Time course (days) Infected = 6 Non-infected controls = 6 (after 30, 60, 90, 120, 150, 180, 210 and 240 days) Infected = 12 Non-infected controls = 12 after 260 days	Brain tissue homogenate	CM10 Q10	Direct comparison of normalized peak intensities + Cluster analysis + PCA	No	No Biomarkers identified: Hsp/Cpn 10, DBI/ACBP and FKBP12	Yes	[3]

Note: See main text for more detailed information and discussion of these studies.

Abbreviations: HCC, hepatocellular carcinoma; LC, liver cirrhosis; CHAPS, 3-[(3-cholamidopropyl)dimethylammonio]-1-propanesulfonate hydrate; HBV, hepatitis B virus; SARS, severe acute respiratory syndrome; CM10, weak cationic exchange array; Q10, strong anionic exchange array; IMAC, immobilised metal affinity array; EBV, Epstein–Barr virus; OSCC, oral squamous cell carcinomas; HPV, human papillomavirus; PRSS, porcine reproductive and respiratory disease; PTB, pulmonary tuberculosis; PTL, preterm labour; AF, amniotic fluid; IAI, intra-amniotic inflammation; PPROM, preterm premature rupture of membranes; ELISA, enzyme linked immunosorbent assay; SAA, serum amyloid A; CF, cystic fibrosis; COPD, chronic obstructive pulmonary disease; BALF, bronchial lavage fluid: IGFBP-1, insulin-like growth factor binding protein-1; TSE, transmissible spongiform encephalopathy, CJD, Creutzfeldt- Jakob disease, PCA, principal component analysis; HAD, HIV-1-associated dementia; CSF, cerebrospinal fluid.

13.10 Concluding Remarks

In our review of the use in SELDI in infectious diseases in 2007 [34], we predicted that the technique would be increasingly used for biomarker identification because of its requirement for only small amounts of sample and high throughput capability. Whilst there have been a number of studies, it is surprising that there has not been more. What are the reasons? Firstly, there was some concern about the level of support that Bio-Rad would give to the SELDI platform after its acquisition from Ciphergen. In the event any such fears were unfounded. Secondly, the method requires expensive equipment and consumables. Therefore, access to SELDI technology and the price of ProteinChips especially for large studies requires significant investment. Thirdly, the accuracy and sensitivity of the SELDI Reader compared with other mass spectrometric techniques such as MALDI-TOF have been questioned [70]. In general, the availability of MALDI-TOF is greater than SELDI and the former is increasingly being used for "SELDI-like" biomarker studies. Partly, this is due to the advent of methods designed for use with MALDI-TOF that mimic the ProteinChip element of SELDI. For example, Invitrogen sell Dynabeads®WCX (weak cation exchange) magnetic particles which can be used for serum profiling. They are a direct equivalent of WCX2 ProteinChips and are used in the same way i.e. the use of different buffer and pH wash conditions results in enrichment of a sub-proteome. Eluates are analysed on a MALDI-TOF machine. Further equivalent products are in development.

Nonetheless, as the above review of studies illustrates, SELDI can be a useful tool in biomarker discovery if suitable care is taken in collection and storage of the samples, sufficient numbers are analysed, steps are taken to eliminate bias such as randomization of samples, operators are blind to the samples they are analysing, and the use of random training and test sets as well as new cohorts for validation. One problem in early studies was that it was comparatively difficult to go from an m/z peak to identification at the molecular level. However, the advances in other peptide/protein methods particularly the use and availability of the ProteomeLab PF2D system (Beckmann Coulter) and liquid chromatography tandem mass spectrometric methods has made identification easier. PF2-D involves two-dimensional liquid separation of protein samples. Samples are separated by chromatofocusing and reverse phase chromatography in the first and second dimensions, respectively. In regard to SELDI, Bio-Rad now produce the Lucid Proteomics System which couples ProteinChip SELDI arrays with a high performance tandem mass spectrometer from Bruker Daltonics. The Lucid Proteomics System should facilitate molecular identity of peaks of interest (biomarkers) obtained from SELDI studies, although it is optimal in identifying peptides < 5 kDa. Another advance/product that may influence investigators to use the SELDI platform in the future is the availability of ProteoMiner beads which consist of a library of combinatorial peptide ligands to simultaneously dilute high-abundance and concentrate low abundance proteins. Availability of ProteoMiner beads in slurry form, such that they can be used in multi-well formats, will potentially allow their use in a high throughput capacity. An alternative method, comparatively widely used in proteomic studies, is immunodepletion of the most abundant proteins in serum which again increases the dynamic range of proteins to be analysed. Thus, less abundant proteins may be amenable to detection. However, immunodepletion kits are usually supplied in cartridge form and are cost-prohibitively expensive if large numbers of samples are to be analysed.

Also in our previous review, we envisaged that SELDI would not become a routine procedure in clinical biochemistry laboratories primarily because of cost. The literature suggests that this prediction has been borne out. The one possible exception is the use in some institutions to detect those at risk from preterm birth [7]. We predict that the status quo will be maintained and SELDI will primarily be used to identify biomarkers for infectious diseases with the aim of developing simpler tests for routine laboratory or, particularly in the case of the developing world, at the point of care. Ndao *et al.* [53] stated that "rather than replacing antibody- and nucleic acid-based testing, it seems more likely that MS (mass spectrometric) or MS-derived assays will provide complementary information (e.g. confirmatory testing, tests for cure or prognosis)". We concur with their view for the developed, although clearly not the developing world.

In summary, SELDI data have identified biomarkers and provided valuable biological insights into host-pathogen interactive biology for a number of infectious diseases. It is likely that the use of SELDI in combination with new developments, such as ProteoMiner beads and methods for rapid protein identification, will result in its continued use for biomarker identification and greater understanding of infectious disease processes.

References

1. Altamura, S., Kiss, J., Blattmann, C., Gilles, W., and Muckenthaler, M.U. (2009). SELDI-TOF MS detection of urinary hepcidin, *Biochimie,* **91**, pp. 1335–1338.

2. Andre, F.E., Booy, R., Bock, H.L., Clemens, J., Datta, S.K., John, T.J., Lee, B.W., Lolekha, S., Peltola, H., Ruff, T.A., Santosham, M., and Schmitt, H.J. (2008). Vaccination greatly reduces disease, disability, death and inequity worldwide, *Bull World Health Organ,* **86**, pp. 140–146.

3. Barr, J.B., Watson, M., Head, M.W., Ironside, J.W., Harris, N., Hogarth, C., Fraser, J.R., and Barron, R. (2009). Differential protein profiling as a potential multi-marker approach for TSE diagnosis, *BMC Infect Dis,* **9**, 188.

4. Bhattacharjee, M., Botting, C.H., and Sillanpää, M.J. (2008). Bayesian biomarker identification based on marker-expression proteomics data, *Genomics,* **6**, pp. 384–392.

5. Blanco, J.D. (2000). Clinical problem of preterm labor, *Clin Obstet Gynecol,* **43**, pp. 713–716.

6. Bozinovski, S., Hutchinson, A., Thompson, M., Macgregor, L., Black, J., Giannakis, E., Karlsson, A.S., Silvestrini, R., Smallwood, D., Vlahos, R., Irving, L.B., and Anderson, G.P. (2008). Serum amyloid a is a biomarker of acute exacerbations of chronic obstructive pulmonary disease, *Am J Respir Crit Care Med,* **177**, pp. 269–278.

7. Buhimschi, C.S., Bhandari, V., Hamar, B.D., Bahtiyar, M.O., Zhao, G., Sfakianaki, A.K., Pettker, C.M., Magloire, L., Funai, E., Norwitz, E.R., Paidas, M., Copel, J.A., Weiner, C.P., Lockwood, C.J., and Buhimschi, I.A. (2007). Proteomic profiling of the amniotic fluid to detect inflammation, infection, and neonatal sepsis, *PLoS Med,* **4,** e18.

8. Buhimschi, C.S., Dulay, A.T., Abdel-Razeq, S., Zhao, G., Lee, S., Hodgson, E.J., Bhandari, V., and Buhimschi, I.A. (2009). Fetal inflammatory response in women with proteomic biomarkers characteristic of itra-amniotic inflammation and preterm birth, *BJOG,* **116**, pp. 257–267.

9. Buhimschi, I.A., Christner, R., and Buhimschi, C.S. (2005). Proteomic biomarker analysis of amniotic fluid for identification of intra-amniotic inflammation, *BJOG,* **112**, pp. 173–181.

10. Buhimschi, I.A., Zambrano, E., Pettker, C.M., Bahtiyar, M.O., Paidas, M., Rosenberg, V.A., Thung, S., Salafia, C.M., and Buhimschi, C.S. (2008).

Using proteomic analysis of the human amniotic fluid to identify histologic chorioamnionitis, *Obstet Gynecol*, **111**, pp. 403–412.

11. Buhimschi, I.A., Zhao, G., Rosenberg, V.A., Abdel-Razeq, S., Thung, S., and Buhimschi, C.S. (2008) Multidimensional proteomics analysis of amniotic fluid to provide insight into the mechanisms of idiopathic preterm birth, *PLoS ONE*, **3**, e2049.
12. Caffrey, R.E. (2010). A Review of experimental design best practices for proteomics based biomarker discovery: Focus on SELDI-TOF. *Methods Mol Biol*, **641**, pp. 167–183.
13. Carrette, O., Burkhard, P.R., Hughes, S., Hochstrasser, D.F., and Sanchez, J.C. (2005). Truncated cystatin C in cerebrospiral fluid: Technical artefact or biological process?, *Proteomics*, **5**, pp. 3060–3065.
14. Castagna, A., Campostrini, N., Zaninotto, F., and Girelli, D. (2010). Hepcidin assay in serum by SELDI-TOF-MS and other approaches, *J Proteomics*, **73**, pp. 527–536.
15. Challis, J.R.G. (2000). Mechanisms of parturition and preterm labor, *Obstet and Gynecol Surv*, **55**, pp. 650–660.
16. Chen, L., Ho, D.W., Lee, N.P., Sun, S., Lam, B., Wong, K.F., Yi, X., Lau, G.K., Ng, E.W., Poon, T.C., Lai, P.B., Cai, Z., Peng, J., Leng, X., Poon, R.T., and Luk, J.M. (2010). Enhanced detection of early hepatocellular carcinoma by serum SELDI-TOF proteomic signature combined with alpha-fetoprotein marker, *Ann Surg Oncol*, **17**, pp. 2518–2525.
17. Cui, J.F., Liu, Y.K., Zhou, H.J., Kang, X.N., Huang, C., He, Y.F., Tang, Z.Y., and Uemura, T. (2008). Screening serum hepatocellular carcinoma-associated proteins by SELDI-based protein spectrum analysis, *World J Gastroenterol*, **14**, pp. 1257–1262.
18. Datta, S., and Pihur, V. (2010). Feature selection and machine learning with mass spectrometry data, *Methods Mol Biol*, **593**, pp. 205–229.
19. De Bock, M., de Seny, D., Meuwis, M.A., Chapelle, J.P., Louis, E., Malaise, M., Merville, M.P., and Fillet, M. (2010). Challenges for biomarker discovery in body fluids using SELDI-TOF-MS, *J Biomed Biotechnol*, 906082. doi:10.1155/2010/906082.
20. Deckers, N., Dorny, P., Kanobana, K., Vercruysse, J., Gonzalez, A.E., Ward, B., and Ndao, M. (2008). Use of ProteinChip technology for identifying biomarkers of parasitic diseases: the example of porcine cysticercosis (*Taenia solium*), *Exp Parasitol*, **120**, pp. 320–329.
21. El-Aneed, A., and Banoub, J. (2006). Proteomics in the diagnosis of hepatocellular carcinoma: focus on high risk hepatitis B and C patients, *Anticancer Res*, **26**, pp. 3293–3300.

22. Erlwein, O., Kaye, S., McClure, M.O., Weber, J., Wills, G., Collier, D., Wessely, S., and Cleare, A. (2010). Failure to Detect the Novel Retrovirus XMRV in Chronic Fatigue Syndrome. *PLoS ONE*, **5**, e8519.

23. Fujita, N., Sugimoto, R., Motonishi, S., Tomosugi, N., Tanaka, H., Takeo, M., Iwasa, M., Kobayashi, Y., Hayashi, H., Kaito, M., and Takei, Y. (2008). Patients with chronic hepatitis C achieving a sustained virological response to peginterferon and ribavirin therapy recover from impaired hepcidin secretion. *J Hepatol*, **49**, pp. 702–710.

24. Fujita, N., Nakanishi, M., Mukai, J., Naito, Y., Ichida, T., Kaito, M., Yoshikawa, T., and Takei, Y. (2010). Identification of treatment efficacy related host factors in chronic hepatitis C by ProteinChip serum analysis, *Mol Med*, **17**, pp. 70–78.

25. Genini, S., Cantu, M., Botti, S., Malinverni, R., Costa, A., Marras, D., and Giuffra, E. (2008). Diagnostic markers for diseases: SELDI-TOF profiling of pig sera for PRRS, *Dev Biol* (Basel), **132**, pp. 399–403.

26. Gomes-Alves, P., Imrie, M., Gray, R.D., Nogueira, P., Ciordia, S., Pacheco, P., Azevedo, P., Lopes, C., de Almeida, A.B., Guardiano, M., Porteous, D.J., Albar, J.P., Boyd, A.C., and Penque, D. (2010). SELDI-TOF biomarker signatures for cystic fibrosis, asthma and chronic obstructive pulmonary disease, *Clin Biochem*, **43**, pp. 168–177.

27. Gray, R.D., MacGregor, G., Noble, D., Imrie, M., Dewar, M., Boyd, A.C., Innes, J.A., Porteous, D.J., and Greening, A.P. (2008). Sputum proteomics in inflammatory and suppurative respiratory diseases, *Am J Respir Crit Care Med*, **178**, pp. 444–452.

28. Hack, C.J. (2004). Integrated transcriptome and proteome data: The challenges ahead, *Brief Funct Genomic Proteomic*, **3**, pp. 212–219.

29. Hamilton, S., Levin, M., Kroll, J.S., and Langford, P. (2010) Microbial disease markers using ProteinChip arrays. In: *Mass Spectroscopy for Microbial Proteomics* (John Wiley and Sons Ltd, UK), pp. 225–250.

30. Han, Y.W., Shen, T., Chung, P., Buhimschi, I.A., and Buhimschi, C.S. (2009). Uncultivated bacteria as etiologic agents of intra-amniotic inflammation leading to preterm birth, *J Clin Microbiol*, **47**, pp. 38–47.

31. Groom, H.C., Boucherit, V.C., Makinson, K., Randal, E., Baptista, S., Hagan, S., Gow, J.W., Mattes, F.M., Breuer, J., Kerr, J.R., Stoye, J.P., and Bishop, K.N. (2010). Absence of xenotropic murine leukaemia virus-related virus in UK patients with chronic fatigue syndrome, *Retrovirology*, **7**, p. 10.

32. He, M., Qin, J., Zhai, R., Wei, X., Wang, Q., Rong, M., Jiang, Z., Huang, Y., and Zhang, Z. (2008). Detection and identification of NAP-2 as a biomarker in hepatitis B-related hepatocellular carcinoma by proteomic approach, *Proteome Sci*, **6**, p. 10.

33. He, Q.Y., Zhu, R., Lei, T., Ng, M.Y., Luk, J.M., Sham, P., Lau, G.K., and Chiu, J.F. (2008). Toward the proteomic identification of biomarkers for the prediction of HBV related hepatocellular carcinoma, *J Cell Biochem*, **103**, pp. 740–752.

34. Hodgetts, A., Levin, M., Kroll, J.S., and Langford, P.R. (2007). Biomarker discovery in infectious diseases using SELDI, *Future Microbiol*, **2**, pp. 35–49.

35. Hu, J., Coombes, K.R., Morris, J.S., and Baggerly, K.A. (2005). The importance of experimental design in proteomic mass spectrometry experiments: Some cautionary tales, *Brief Funct Genomic Proteomic*, **3**, pp. 322–331.

36. Huang, Y.J., Xuan, C., Zhang, B.B., Liao, M., Deng, K.F., He, M., and Zhao, J.M. (2009). SELDI-TOF MS profiling of serum for detection of nasopharyngeal carcinoma, *J Exp Clin Cancer Res*, **28**, 85.

37. Huang, Y.W., and Meng, X.J. (2010). Novel strategies and approaches to develop the next generation of vaccines against porcine reproductive and respiratory syndrome virus (PRRSV), *Virus Res*, **154**(1–2), 141–149.

38. Jones, K.E., Patel, N.G., Levy, M.A., Storeygard, A., Balk, D., Gittleman, J.L. and Daszak, P. (2008). Global trends in emerging infectious diseases, *Nature*, **451**, pp. 990–993.

39. Kanmura, S., Uto, H., Sato, Y., Kumagai, K., Sasaki, F., Moriuchi, A., Oketani, M., Ido, A., Nagata, K., Hayashi, K., Stuver, S.O., and Tsubouchi, H. (2010). The complement component C3a fragment is a potential biomarker for hepatitis C virus-related hepatocellular carcinoma, *J Gastroenterol*, **45**, pp. 459–467.

40. Kerr, J.R., Christian, P., Hodgetts, A., Langford, P.R., Devanur, L.D., Petty, R., Burke, B., Sinclair, L.I., Richards, S.C., Montgomery, J., McDermott, C.R., Harrison, T.J., Kellam, P., Nutt, D.J., and Holgate, S.T. (2007). Collaborative Clinical Study Group.Current research priorities in chronic fatigue syndrome/myalgic encephalomyelitis: disease mechanisms, a diagnostic test and specific treatments, *J Clin Pathol*, **60**, pp. 113–116.

41. Kerr, J.R., Petty, R., Burke, B., Gough, J., Fear, D., Sinclair, L.I., Mattey, D.L., Richards, S.C., Montgomery, J., Baldwin, D.A., Kellam, P., Harrison, T.J., Griffin, G.E., Main, J., Enlander, D., Nutt, D.J., and Holgate, S.T. (2008). Gene expression subtypes in patients with chronic fatigue syndrome/myalgic encephalomyelitis, *J Infect Dis*, **197**, pp. 1171–1184.

42. Kiehntopf, M., Siegmund, R. and Deufel, T. (2007). Use of SELDI-TOF mass spectrometry for identification of new biomarkers: potential and limitations, *Clin Chem Lab Med,* **45**, pp. 1435–1449.

43. Lee, I.N., Chen, C.H., Sheu, J.C., Lee, H.S., Huang, G.T., Chen, D.S., Yu, C.Y., Wen, C.L., Lu, F.J. and Chow, L.P. (2006). Identification of complement C3a as a candidate biomarker in human chronic hepatitis C and HCV-related hepatocellular carcinoma using a proteomics approach, *Proteomics*, **6**, pp. 2865–2873.

44. Lin, W.T., Tsai, C.C., Chen, C.Y., Lee, W.J., Su, C.C. and Wu, Y.J. (2008). Proteomic analysis of peritoneal dialysate fluid in patients with dialysis-related peritonitis, *Ren Fail*, **30**, pp. 772–777.

45. Liu, Q., Chen, X., Hu, C., Zhang, R., Yue, J., Wu, G., Li, X., Wu, Y. and Wen, F. (2010). Serum protein profiling of smear-positive and smear-negative pulmonary tuberculosis using SELDI-TOF mass spectrometry, *Lung*, **188**, pp. 15–23.

46. Liu, Z., Ma, Y., Yang, J. and Qin, H. (2010). Upregulated and downregulated proteins in hepatocellular carcinoma: A systematic review of proteomic profiling Studies, *OMICS,* **15**(1–2), pp. 61–71.

47. Llovet, J.M., Burroughs, A., and Bruix, J. (2003). Hepatocellular carcinoma. *Lancet,* **362**, pp. 1907–1917.

48. Lo, S.C., Pripuzova, N., Li, B., Komaroff, A.L., Hung, G.C., Wang, R., and Alter, H.J. (2010). Detection of MLV-related virus gene sequences in blood of patients with chronic fatigue syndrome and healthy blood donors, *Proc Natl Acad Sci U S A*, **107**, pp. 15874–15879.

49. Lombardi, V.C., Ruscetti, F.W., Das Gupta, J., Pfost, M.A., Hagen, K.S., Peterson, D.L., Ruscetti, S.K., Bagni, R.K., Petrow-Sadowski, C., Gold, B., Dean, M., Silverman, R.H., and Mikovits, J.A. (2009). Detection of an infectious retrovirus, XMRV, in blood cells of patients with chronic fatigue syndrome. *Science*, **326**, pp. 585–589.

50. MacGregor, G., Gray, R.D., Hilliard, T.N., Imrie, M., Boyd, A.C., Alton, E.W., Bush, A., Davies, J.C., Innes, J.A., Porteous, D.J., and Greening, A.P. (2008). Biomarkers for cystic fibrosis lung disease: application of SELDI-TOF mass spectrometry to BAL fluid, *J Cyst Fibros*, **7**, pp. 352–358.

51. McMorran, B.J., Patat, S.A., Carlin, J.B., Grimwood, K., Jones, A., Armstrong, D.S., Galati, J.C., Cooper, P.J., Byrnes, C.A., Francis, P.W., Robertson, C.F., Hume, D.A., Borchers, C.H., Wainwright, C.E., and Wainwright, B.J. (2007) Novel neutrophil-derived proteins in bronchoalveolar lavage fluid indicate an exaggerated inflammatory response in pediatric cystic fibrosis patients, *Clin Chem*, **53**, pp. 1782–17891.

52. Melle, C., Ernst, G., Winkler, R., Schimmel, B., Klussmann, J.P., Wittekindt, C., Guntinas-Lichius, O., and von Eggeling, F. (2009). Proteomic analysis of human papillomavirus-related oral squamous cell carcinoma: Identification of thioredoxin and epidermal-fatty acid binding protein as upregulated protein markers in microdissected tumor tissue, *Proteomics*, **9**, pp. 2193–21201.

53. Ndao, M., Spithill, T.W., Caffrey, R., Li, H., Podust, V.N., Perichon, R., Santamaria, C., Ache, A., Duncan, M., Powell, M.R., and Ward, B.J. (2010) Identification of novel diagnostic serum biomarkers for Chagas' disease in asymptomatic subjects by mass spectrometric profiling, *J Clin Microbiol*, **48**, pp. 1139–1149.

54. Paugam-Burtz, C., Albuquerque, M., Baron, G., Bert, F., Voitot, H., Delefosse, D., Donderom, F., Sommacale, D., Francoz, C., Hanna, N., Belghiti, J., Ravaud, P., Bedossa, P., Mantz, J., and Paradis, V. (2010). Plasma proteome to look for diagnostic biomarkers of early bacterial sepsis after liver transplantation: a preliminary study, *Anesthesiology*, **112**, pp. 926–935.

55. Ratto-Kim, S., Chuenchitra, T., Pulliam, L., Paris, R., Sukwit, S., Gongwon, S., Sithinamsuwan, P., Nidhinandana, S., Thitivichianlert, S., Shiramizu, B.T., de Souza, M.S., Chitpatima, S.T., Sun, B., Rempel, H., Nitayaphan, S., Williams, K., Kim, J.H., Shikuma, C.M., and Valcour, V.G. (2008). Southeast Asia Research Collaboration with the University of Hawaii (SEARCH) 001 protocol team.Expression of monocyte markers in HIV-1 infected individuals with or without HIV associated dementia and normal controls in Bangkok Thailand, *J Neuroimmunol*, **195**, pp. 100–107.

56. Rioux, M.C., Carmona, C., Acosta, D., Ward, B., Ndao, M., Gibbs, B.F., Bennett, H.P., and Spithill, T.W. (2008). Discovery and validation of serum biomarkers expressed over the first twelve weeks of *Fasciola hepatica* infection in sheep, *Int J Parasitol*, **38**, pp. 123–136.

57. Romero, R., Espinoza, J., Rogers, W.T., Moser, A., Nien, J.K., Kusanovic, J.P., Gotsch, F., Erez, O., Gomez, R., Edwin, S., and Hassan, S.S. (2008). Proteomic analysis of amniotic fluid to identify women with preterm labor and intra-amniotic inflammation/infection: the use of a novel computational method to analyze mass spectrometric profiling, *J Matern Fetal Neonatal Med*, **21**, pp. 367–388.

58. Sánchez, M.I., Thomas, F., Perrot-Minnot, M.J., Biron, D.G., Bertrand-Michel, J., and Missé, D. (2009). Neurological and physiological disorders in Artemia harboring manipulative cestodes, *J Parasitol*, **95**, pp. 20–24.

59. Schwegler, E.E., Cazares, L., Steel, L.F., Adam, B.L., Johnson, D.A., Semmes, O.J., Block, T.M., Marrero, J.A., and Drake, R.R. (2005). SELDI-TOF MS profiling of serum for detection of the progression of chronic hepatitis C to hepatocellular carcinoma, *Hepatology,* **41**, pp. 634–642.

60. Shikuma, C.M., and Valcour, V.G. (2008). Expression of monocyte markers in HIV-1 infected individuals with or without HIV associated dementia and normal controls in Bangkok, Thailand, *J Neuroimmunol,* **195**, pp. 100–107.

61. Silberring, J., and Ciborowski, P. (2010). Biomarker discovery and clinical proteomics, *Trends Analyt Chem,* **29**, 128.

62. Steel, L.F., Shumpert, D., Trotter, M., Seeholzer, S.H., Evans, A.A., London, W.T., Dwek, R., and Block, T.M. (2003). A strategy for the comparative analysis of serum proteomes for the discovery of biomarkers for hepatocellular carcinoma, *Proteomics,* **3**, pp. 601–609.

63. Steinacker, P., Rist, W., Swiatek-de-Lange, M., Lehnert, S., Jesse, S., Pabst, A., Tumani, H., von Arnim, C.A., Mitrova, E., Kretzschmar, H.A., Lenter, M., Wiltfang, J., and Otto, M. (2010). Ubiquitin as potential cerebrospinal fluid marker of Creutzfeldt-Jakob disease, *Proteomics,* **10**, pp. 81–89.

64. Sugimoto, R., Fujita, N., Tomosugi, N., Hara, N., Miyachi, H., Tanaka, H., Takeo, M., Nakagawa, N., Iwasa, M., Kobayashi, Y., Kaito, M., and Takei, Y. (2009). Impaired regulation of serum hepcidin during phlebotomy in patients with chronic hepatitis C, *Hepatol Res,* **9**, pp. 619–624.

65. Sun, S., Lee, N.P., Poon, R.T., Fan, S.T., He, Q.Y., Lau, G.K., and Luk, J.M. (2007). Oncoproteomics of hepatocellular carcinoma: from cancer markers' discovery to functional pathways, *Liver Int,* **27**, pp. 1021–1038.

66. Switzer, W. M., Jia, H., Hohn, O., Zheng, H., Tang, S., Shankar, A., Bannert, N., Simmons, G., Hendry, R.M., Falkenberg, V.R., Reeves, W.C., and Heneine, W. (2010). Absence of evidence of Xenotropic Murine Leukemia Virus-related virus infection in persons with Chronic Fatigue Syndrome and healthy controls in the United States, *Retrovirology,* **7**, 57.

67. van Kuppeveld, F.J., Jong, A.S., Lanke, K.H., Verhaegh, G.W., Melchers, W.J., Swanink, C.M., Bleijenberg, G., Netea, M.G., Galama, J.M., and van der Meer, J.W. (2010). Prevalence of xenotropic murine leukaemia virus-related virus in patients with chronic fatigue syndrome in the Netherlands: Retrospective analysis of samples from an established cohort, *BMJ,* **340**, c1018.

68. Vivekanandan, P., and Singh, O.V. (2008). High-dimensional biology to comprehend hepatocellular carcinoma, *Expert Rev Proteomics,* **5**, pp. 45–60.

69. Wiederin, J., Rozek, W., Duan, F., and Ciborowski, P. (2009). Biomarkers of HIV-1 associated dementia: proteomic investigation of sera. *Proteome Sci*, **7**, 8.

70. Whelan, L.C., Power, K.A., McDowell, D.T., Kennedy, J., and Gallagher, W.M. (2008). Applications of SELDI-MS technology in oncology, *J Cell Mol Med*, **12**, pp. 1535–1547.

71. Wright, L.M., Kreikemeier, J.T., and Fimmel, C.J. (2007). A concise review of serum markers for hepatocellular cancer, *Cancer Detect Prev*, **31**, pp. 35–44.

72. Wu, C., Wang, Z., Liu, L., Zhao, P., Wang, W., Yao, D., Shi, B., Lu, J., Liao, P., Yang, Y., and Zhu, L. (2008). Surface enhanced laser desorption/ionization profiling: New diagnostic method of HBV-related hepatocellular carcinoma, *J Gastroenterol Hepatol*, **24**, pp. 55–62.

73. Wu, F.X., Wang, Q., Zhang, Z.M., Huang, S., Yuan, W.P., Liu, J.Y., Ban, K.C., and Zhao, Y.N. (2009). Identifying serological biomarkers of hepatocellular carcinoma using surface-enhanced laser desorption/ionization-time-of-flight mass spectroscopy, *Cancer Lett*, **279**, pp. 163–170.

74. Zhang, J., Li, D., Zheng, Y., Cui, Y., Feng, K., Zhou, J., and Wu, J. (2008). Proteomic profiling of Hepatitis B virus-related hepatocellular carcinoma in China: A SELDI-TOF-MS Study, *Int J Clin Exp Pathol*, **1**, pp. 352–361.

75. Zinkin, N.T., Grall, F., Bhaskar, K., Otu, H.H., Spentzos, D., Kalmowitz, B., Wells, M., Guerrero, M., Asaram J.M., Libermann, T.A., and Afdhal, N.H. (2008). Serum proteomics and biomarkers in hepatocellular carcinoma and chronic liver disease, *Clin Cancer Res*, **14**, pp. 470–477.

Chapter 14

Quantitative Proteomic Analysis of HIV Infection

Li Juan Fu and Sam F. Y. Li
Department of Chemistry, National University of Singapore, 3 Science Drive 3, Singapore 117543, Singapore
chmlifys@nus.edu.sg

14.1 Introduction

Infection of human immunodeficiency virus type (HIV) 1 in an infected individual affects many physiological processes, eventually leading to the development of acquired immune deficiency syndrome (AIDS). For efficient replication, the virus utilizes its accessory proteins to modify many physiological aspects of the infected cells and the individual as a whole.[1–3] Immense studies have been carried out to study HIV infection such as the role of HIV accessory proteins in the intracellular processes affected by HIV infection and mRNA changes in the infected cells.[4–7] However, it is only recently that the proteome changes are being studied.[8]

Proteomics can be defined as "the systematic study of the many and diverse properties of proteins in a parallel manner with the aim of providing detailed description of the structure, function

Quantitative Proteome Analysis: Methods and Applications
Edited by Kazuhiro Imai and Sam F. Y. Li

ISBN 978-981-4316-51-4 (Hardcover), 978-981-4316-98-9 (eBook)
www.panstanford.com

and control of biological systems in health and disease".[9] The most valuable information on the system being studied is the differential expression of proteins in cells, tissues or body fluids of different states. Hence, proteomic analysis needs to be quantitative. Some of the most common approaches employed in quantitative proteomics are introduced in the following section and the principle associated with each approach is briefly discussed.

Various quantitative proteomic methods have been employed in the study of HIV infection. In the rest of this chapter, examples of HIV research utilizing quantitative proteomic analysis will be provided. As the research is much diversified, the examples are grouped according to the sample type. The proteome of T cells can be monitored for changes in response to HIV infection. The cultured HIV-infected macrophages can be analyzed, or the plasma of infected individuals can be used to study whole body responses to HIV infection. Cerebrospinal fluids from infected patients can be analyzed.

14.2 HIV Proteomic Research

14.2.1 T Cells

In the pathogenesis of HIV, the cell types T cells, monocytes, macrophages, and follicular dendritic cells are primarily responsible for the replication of HIV in the body. Hence, it is logical to study the proteome of these cells. The envelope and cell surface membranes of the virus are made up of lipoproteins and cholesterol and the fatty acids are synthesized in the cytoplasm. In the virus replication starting from its entry into the cell to integration, transcription, assembly and budding of HIV from cells, both lipids and lipoproteins are necessary for molecular communication in each replication step.[10–12] S. Rasheed *et al.* investigated if HIV infection affected the production of any proteins involved in the lipid metabolism in a T-cell line (RH9) *in vitro*.[13] In their study, proteins were extracted from cells of both infected and uninfected cultures. The samples were analyzed by isoelectric focusing (IEF), quantified by 2DE and identified using matrix assisted laser desorption-time of flight (MALDI-TOF) mass spectrometry. Eighteen proteins were identified to be differentially expressed in HIV-infected cells—12 proteins were expressed exclusively, 2 were upregulated, and 4 were downregulated. Out of these 18 proteins, seven were from the various families of enzymes/kinases, five were transporter proteins, two were molecular chaperons, two were transmembrane receptors, and one was a ligand-binding protein and was an adapter-like protein. The study provided the first direct evidence that HIV infection alters the production of proteins that are involved in lipid synthesis, transport and metabolism.

In another study that focused on membrane-associated proteins on HIV-infected T cells, isolated proteins were separated by 2DE and later identified by MALDI-TOF mass spectrometer.[14] Berro *et al.* found that 17 proteins on the surface of T cells were differentially expressed due to HIV infection. Of these proteins, 65% are integral membrane or membrane-associated proteins that could be divided into two functions: (1) proteins involved in cell adhesion, structure, and migration; (2) receptors or receptor-associated proteins involved in the regulation of cell death and survival. Specifically, a protein X-linked inhibitor of apoptosis was identified to be upregulated in HIV-infected T cells. The authors suggested that the upregulation of this protein was a mechanism to increase cell survival to counterbalance the apoptotic effects of some viral accessory proteins.

The proteome response of the human $CD4^+$ cell line CEMx174 to HIV infection was analyzed by quantitative proteomics.[8] Relative quantitative analysis was carried out by utilizing the $^{16}O/^{18}O$ stable isotope labeling[15] and the peptides were analyzed by using electrospray ionization interface-Fourier transform ion cyclotron resonance (FTICR) mass spectrometer. Chan *et al.* quantified 3255 proteins, out of which 344 increased and 343 decreased in abundance. These 687 proteins were mostly involved in certain biological pathways such as ubiquitin-conjugating enzymes in ubiquitination, carrier proteins in nucleocytoplasmic transport, cyclin-dependent kinase in cell cycle, and pyruvate dehydrogenase of the citrate cycle pathways. This indicates that specific cellular processes may be involved in the virus replication process.

CD4 is an integral membrane protein consisting of four extracellular domains, a transmembrane domain, and a cytoplasmic tail.[16] Its function is as a coreceptor in the activation of $CD4^+$ cells[17] and the receptor for the entry of HIV when combined with CCR-5 and CXCR-4.[18,19] Membrane proteins are usually very difficult to study and their low abundance requires additional strategies for their detection. Bernhard *et al.* employed isotopically coded affinity tag (ICAT) to study CD4-associating proteins. They used a cysteinyl affinity capture method to reduce the complex protein tryptic mixture.[20] This limited the mixture to be reduced to only cysteine-containing peptides. CD4 receptor complex components, i.e., CD4 and p56lck and its associating molecules, and several other proteins were identified using this strategy. However, the reduction of protein mixture resulted in only 10% of the digested peptides to be identified. Hence, this strategy did not enable the identification of all proteins.

14.2.2 Macrophages

Macrophages are mononuclear phagocytes that are involved in both inane and adaptive immune responses. They act as protector of the immune systems due to their phagocytic and inflammatory properties. It has been discovered in the late 1980s that HIV can infect monocytes and macrophages.[21–23] HIV persists in the host system even though antiretroviral treatment is administered. There are two theories of this persistence nature of HIV—latency and ongoing replication. Ongoing replication is a consequence of drug resistance, whereas latency involves the presence of HIV in reservoir cells, which include resting memory $CD4^+$ T cells and mononuclear phagocytes such as peripheral blood monocytes, macrophages, microglia, and dendritic cells.[24–32] Hence, new therapies are required to purge these reservoir cells in order for effective treatment. Since macrophages belong to one of these reservoirs, it makes sense to study how HIV affects macrophages.

HIV-infected individuals exhibiting a higher risk of invasive bacterial infections although highly active antiretroviral therapy (HAART) was administered and $CD4^+$ T cell counts had recovered.[33,34] The mechanisms behind this are not fully understood yet. However, some studies suggest that when HIV infects key cells of the monocyte-macrophage lineage, the innate immune system to bacteria is impaired. Thus, Pathak *et al.* studied the effects of HIV infection on the proteome of undifferentiated monocyte-like THP-1 cells.[35] SILAC was employed in this study and 651 proteins were identified. Out of these 651 proteins, 9 proteins were downregulated and 17 were upregulated in HIV-infected THP-1 cells. It was found that the IL-1 receptor-associated kinase 4 (IRAK-4) was downregulated. IRAK-4 is essential for virtually all Toll-like receptors (TLR) signaling cascade for pathogen recognition in monocytes/macrophages. Hence, it is important in effective immune response of the innate and adaptive immune system. The authors, thus, suggested that the ability of the innate immune system to sense pathogens may be diminished because HIV may impair the TLR signaling cascade for pathogen recognition.

In a study to determine the role regulatory T cells (Treg) play in HIV disease, quantitative proteomic analysis utilizing iTRAQ was used to study the differential protein expression of HIV-infected bone marrow-derived macrophages (BMM) with Treg and without Treg[36]; 3553 proteins were identified. The alternation of the proteome of HIV-infected bone marrow-derived macrophages treated with Treg was associated with antiviral response and apoptosis. Interferon-stimulated gene 15 was upregulated. Interferon-stimulated gene 15 is an ubiquitin-like protein that is involved in interferon-mediated antiviral immunity. Hence, it was also observed that with the upregulation of this protein, the virus release was reduced.

14.2.3 Cerebrospinal Fluid

HIV infection affects the development of cognitive, behavioral, and motor deficits and in the most severe form, HIV-associated cognitive impairment is called HIV-associated dementia (HAD).[37] HAD can progress to extreme functional disability and death. Currently, the diagnosis of HAD is made primarily on the basis of the elimination of concurrent opportunistic infections, malignancies, and psychiatric disorders. Hence, there are a number of studies to uncover biomarkers for HAD and also for therapeutic monitoring.[38,39]

Cerebrospinal fluid (CSF) provides nutrients, protection, and environment support for the nervous system.[40] It is generally accepted that CSF proteins reflect the central nervous system (CNS) metabolic activity and, hence, can indicate brain pathological processes.[41,42] Therefore proteomic analysis of CSF can provide information on CNS in diseases.[43–47] In a comprehensive proteomic analysis on CSF of HIV-infected individuals using surface enhanced laser desorption ionization time of-flight (SELDI-TOF), nine proteins were identified to be unique to HIV-associated cognitive impairment. They were soluble superoxide dismutase (SOD 1), related to amyotrophic lateral sclerosis (ALS), migration inhibitory factor (MIF)-related protein 14, macrophage capping protein, neurosecretory protein VGF, galectin-7, L-plastin, acylphosphatase 1, and a tyrosine 3/tryptophan 5-monooxygenase activation protein. These proteins are involved in cell signaling, structural function and antioxidant activities.[48] Rozek *et al.* utilized 2D-DIGE to discover potential biomarkers for HAD.[49] They had discovered more than 90 differentially expressed proteins and identified 20 of them. The differential expression of six proteins—vitamin D binding protein, clusterin, gelsolin, complement C3, procollagen C-endopeptidase enhancer 1, and cystatin C—was validated by Western blotting. The authors, hence, suggested that these proteins were potential biomarkers for HAD.

14.2.4 Plasma

The proteome of plasma has become easier to profile with the development of proteomic tools, and it has been proven to be useful in diagnostic, prognostic and/or therapeutic development.[50] In the objectives of many proteomic studies, the analysis of proteins associated with HIV-infection could potentially identify biomarkers and drug targets.[51,52] In one such study, 2DE was used to analyze HIV-infected plasma samples and non-infected plasma samples collected from infected patients and healthy individuals, respectively.[53] At least eight proteins were expressed differentially in HIV-infected plasma. In particular, apolipoprotein AI was found to be expressed differently for different isoforms. This was the first time such phenomenon was observed in HIV infection. The difference in expression levels reflects that there are different post-translational modifications occurring for different isoforms of apolipoprotein AI. The authors, therefore, suggested that this heterogeneous alteration may provide useful diagnostic information for HIV infection and therapy.

In 2008, Rozek *et al.* posited the sera as an alternative to biomarker detection for cognitive impairment. This study was the first attempt to find HIV-1-associated biomarkers in sera. Eighty proteins spots were differentially expressed, but only 17 spots were identified with high confidence. Only two proteins were validated with Western blotting, namely afamin and ceruloplasmin. Although the roles of these proteins in the sera of patients with HAD are not clear, they may be used as potential biomarkers and the combination of laboratory tests using CSF and sera as samples may aid in the future diagnosis of neurodegenerative diseases.[54]

Recently, another study evaluated the proteomic profile of plasma samples from HIV/HCV mono- and coinfected patients to investigate the changes in protein abundance and to understand how these proteins are involved in the pathophysiology of liver disease in these patients.[55] The plasma samples were subjected to 4-plex iTRAQ labeling and LC-MS/MS for quantification and identification. Seventy proteins were identified with high confidence, out of which apolipoproteins and complement proteins are the two major classes that were differentially expressed. Upregulated proteins include apolipoprotein A-II (APOA2), apolipoprotein C-II (APOC2), apolipoprotein E (APOE), Complement C3 (C3), and histidine-rich glycoprotein (HRG). These proteins play a role in the hepatic lipid metabolism, inflammation and acute-phase response signaling pathways. Thus, these proteins can potentially be biomarkers for liver disease.

14.2.5 Cervical Lavage Samples

Sex workers who are highly exposed to HIV infection but yet remain uninfected can be characterized to be relatively resistant to HIV infection. Evidence has shown that some innate immune factors in the mucosal layer in the cervicovaginal compartment may aid in resistance against HIV infection.[56–60] Burgener *et al.* used 2D-DIGE to do a more thorough characterization of the differences in genital secretions between HIV-resistant, HIV-infected, and HIV-uninfected individuals.[61] More than 15 proteins were found to be differentially expressed between the HIV-resistant group and the control group. Proteins such as antiproteases, including those from the serpin B family and cystatin, a known anti-HIV factor, were found to be highly expressed in HIV-resistant women. Western blotting confirmed the trends observed in 2D-DIGE for some of the proteins. By understanding how these factors play a role in HIV resistance, microbicides and/or vaccines against HIV may be developed.

14.3 Discussion and Conclusion

In HIV infection proteomic research reviewed herein, descriptive studies have pointed to the discovery of biomarkers, infection mechanism, and protection mechanism against HIV infection with the aid of the various proteomic techniques. These methods have helped identify the proteins and quantify their relative amounts in experimental and control samples.

Melendez *et al.*[62] reviewed the applications of proteomics in HIV-infected macrophages. The advantages and limitations of each technique were described, and this suggests that there is not a single quantitative proteomic technique that can provide a complete picture of the biological system under investigation. Some comparative studies of proteomic methods have concluded that the different methods can provide complementary information[63,64] and, hence, can provide a better picture of the biological system under investigation.

In summary, quantitative proteomics is very useful in the research of HIV infection, but it is more efficient when a variety of techniques can be employed in the research to give a more complete picture of HIV infection in the host.

References

1. Perez, O.D., and Nolan, G.P. (2001) Resistance is futile: assimilation of cellular machinery by HIV-1. *Immunity* **15**, 687–690.
2. Gomez, C., and Hope, T.J. (2005) The ins and outs of HIV replication. *Cellular Microbiology* **7**, 621–626.
3. Sierra, S., Kupfer, B., and Kaiser, R. (2005) Basics of the virology of HIV-1 and its replication. *Journal of Clinical Virology* **34**, 233–244.
4. Unwin, R.D., Evans, C.A., and Whetton, A.D. (2006) Relative quantification in proteomics: new approaches for biochemistry. *Trends in Biochemical Sciences* **31**, 473–484.
5. Roeth, J.F., and Collins, K.L. (2006) Human immunodeficiency virus type 1 Nef: adapting to intracellular trafficking pathways. *Microbiology and Molecular Biology Reviews* **70**, 548–563.
6. Giri, M.S., Nebozhyn, M., Showe, L., and Montaner, L.J. (2006) Microarray data on gene modulation by HIV-1 in immune cells: 2000–2006. *Journal of Leukocyte Biology* **80**, 1031–1043.
7. Coiras, M., *et al.* (2006) Modifications in the human T cell proteome induced by intracellular HIV-1 Tat protein expression. *Proteomics* **6**, S63–S73.

8. Chan, E.Y., *et al.* (2007) Quantitative analysis of human immunodeficiency virus type 1-infected CD4$^+$ cell proteome: dysregulated cell cycle progression and nuclear transport coincide with robust virus production. *Journal of Virology* **81**, 7571–7583.

9. Patterson, S.D., and Aebersold, R.H. (2003) Proteomics: the first decade and beyond. *Nature Genetics* **33**, 311–323.

10. Vidalain, P.O., *et al.* (2000) CD40 signaling in human dendritic cells is initiated within membrane rafts. *The EMBO Journal* **19**, 3304–3313.

11. Raulin, J. (2002) Human immunodeficiency virus and host cell lipids. Interesting pathways in research for a new HIV therapy. *Progress in Lipid Research* **41**, 27–65.

12. Liao, Z., Cimakasky, L.M., Hampton, R., Nguyen, D.H., and Hildreth, J.E.K. (2001) Lipid rafts and HIV pathogenesis: host membrane cholesterol is required for infection by HIV type 1. *AIDS Research and Human Retroviruses* **17**, 1009–1019.

13. Rasheed, S., Yan, J.S., Lau, A., and Chan, A.S. (2008) HIV replication enhances production of free fatty acids, low density lipoproteins and many key proteins involved in lipid metabolism: a proteomics study. *PLoS One* **3**, e3003.

14. Berro, R., *et al.* (2007) Identifying the membrane proteome of HIV-1 latently infected cells. *Journal of Biological Chemistry* **282**, 8207–8218.

15. Qian, W.J., *et al.* (2005) Quantitative proteome analysis of human plasma following *in vivo* lipopolysaccharide administration using $^{16}O/^{18}O$ labeling and the accurate mass and time tag approach. *Molecular and Cellular Proteomics* **4**, 700–709.

16. Maddon, P.J., *et al.* (1987) Structure and expression of the human and mouse T4 genes. *Proceedings of the National Academy of Sciences* **84**, 9155–9159

17. Gay, D., *et al.* (1987) Functional interaction between human T-cell protein CD4 and the major histocompatibility complex HLA-DR antigen. *Nature* **328**, 626–629.

18. Wu, L., *et al.* (1996) CD4-induced interaction of primary HIV-1 gp120 glycoproteins with the chemokine receptor CCR-5. *Nature* **384**, 179–183.

19. Lapham, C.K., *et al.* (1996) Evidence for cell-surface association between fusin and the CD4-gp120 complex in human cell lines. *Science* **274**, 602–605.

20. Bernhard, O.K., Burgess, J.A., Hochgrebe, T., Sheil, M.M., and Cunningham, A.L. (2003) Mass spectrometry analysis of CD4-associating proteins using affinity chromatography and affinity tag-mediated purification of tryptic peptides. *Proteomics* **3**, 139–146.
21. Cheng-Mayer, C., Homsy, J., Evans, L., and Levy, J. (1988) Identification of human immunodeficiency virus subtypes with distinct patterns of sensitivity to serum neutralization. *Proceedings of the National Academy of Sciences* **85**, 2815–2819.
22. Gendelman, H.E., *et al.* (1988) Efficient isolation and propagation of human immunodeficiency virus on recombinant colony-stimulating factor 1-treated monocytes. *The Journal of Experimental Medicine* **167**, 1428–1441.
23. Collman, R., *et al.* (1989) Infection of monocyte-derived macrophages with human immunodeficiency virus type 1 (HIV-1). Monocyte-tropic and lymphocyte-tropic strains of HIV-1 show distinctive patterns of replication in a panel of cell types. *The Journal of Experimental Medicine* **170**, 1149–1163.
24. Le Douce, V., Herbein, G., Rohr, O., and Schwartz, C. (2010) Review molecular mechanisms of HIV-1 persistence in the monocyte-macrophage lineage Retrovirology 7: 32 doi:10.1186/1742-4690-7-32.
25. Embretson, J., *et al.* (1993) Analysis of human immunodeficiency virus-infected tissues by amplification and *in situ* hybridization reveals latent and permissive infections at single-cell resolution. *Proceedings of the National Academy of Sciences* **90**, 357–361.
26. Finzi, D., *et al.* (1997) Identification of a reservoir for HIV-1 in patients on highly active antiretroviral therapy. *Science* **278**, 1295–1300.
27. Chun, T.W., Chadwick, K., Margolick, J., and Siliciano, R.F. (1997) Differential susceptibility of naive and memory $CD4^+$ T cells to the cytopathic effects of infection with human immunodeficiency virus type 1 strain LAI. *Journal of Virology* **71**, 4436–4444.
28. Wong, J.K., *et al.* (1997) Reduction of HIV-1 in blood and lymph nodes following potent antiretroviral therapy and the virologic correlates of treatment failure. *Proceedings of the National Academy of Sciences* **94**, 12574–12579
29. Bailey, J.R., *et al.* (2006) Residual human immunodeficiency virus type 1 viremia in some patients on antiretroviral therapy is dominated by a small number of invariant clones rarely found in circulating $CD4^+$ T cells. *Journal of virology* **80**, 6441–6457.

30. Keele, B.F., *et al.* (2008) Identification and characterization of transmitted and early founder virus envelopes in primary HIV-1 infection. *Proceedings of the National Academy of Sciences* **105**, 7552–7557.

31. Zhu, T. (2002) HIV-1 in peripheral blood monocytes: an underrated viral source. *Journal of Antimicrobial Chemotherapy* **50**, 309–311.

32. Zhu, T., *et al.* (2002) Evidence for human immunodeficiency virus type 1 replication *in vivo* in $CD14^+$ monocytes and its potential role as a source of virus in patients on highly active antiretroviral therapy. *Journal of Virology* **76**, 707–716.

33. Noursadeghi, M., Katz, D.R., and Miller, R.F. (2006) HIV-1 infection of mononuclear phagocytic cells: the case for bacterial innate immune deficiency in AIDS. *The Lancet Infectious Diseases* **6**, 794–804.

34. Kaplan, J.E., *et al.* (2000) Epidemiology of human immunodeficiency virus-associated opportunistic infections in the United States in the era of highly active antiretroviral therapy. *Clinical Infectious Diseases* **30**, S5–S14.

35. Pathak, S., De Souza, G., Salte, T., Wiker, H., and Asjo, B. (2009) HIV induces both a down-regulation of IRAK-4 that impairs TLR signalling and an Up-regulation of the antibiotic peptide dermcidin in monocytic cells. *Scandinavian Journal of Immunology* **70**, 264–276.

36. Huang, X., Stone, D.K., Yu, F., Zeng, Y., and Gendelman, H.E. (2010) Functional proteomic analysis for regulatory T cell surveillance of the HIV-1-infected macrophage. *Journal of Proteome Research* **9**, 6759–6773.

37. Ances, B.M., and Ellis, R.J. (2007) Dementia and neurocognitive disorders due to HIV-1 infection. *Seminars in Neurology* **27**, 86–92.

38. Ciborowski, P., and Gendelman, H.E. (2006) Human immunodeficiency virus-mononuclear phagocyte interactions: emerging avenues of biomarker discovery, modes of viral persistence and disease pathogenesis. *Current HIV Research* **4**, 279–291.

39. Sacktor, N., *et al.* (2004) Novel markers of oxidative stress in actively progressive HIV dementia. *Journal of Neuroimmunology* **157**, 176–184.

40. Eugenin, E., *et al.* (2006) Shedding of PECAM-1 during HIV infection: a potential role for soluble PECAM-1 in the pathogenesis of NeuroAIDS. *Journal of Leukocyte Biology* **79**, 444–452.

41. Simonsen, A., *et al.* (2008) Identification of a novel panel of cerebrospinal fluid biomarkers for Alzheimer's disease. *Neurobiology of Aging* **29**, 961–968.

42. Simonsen, A.H., *et al.* (2007) Novel panel of cerebrospinal fluid biomarkers for the prediction of progression to Alzheimer dementia in patients with mild cognitive impairment. *Archives of Neurology* **64**, 366–370.

43. Huang, J., *et al.* (2006) Disease biomarkers in cerebrospinal fluid of patients with first-onset psychosis. *PLoS Med* **3**, e428.

44. Fonteh, A.N., *et al.* (2006) Identification of disease markers in human cerebrospinal fluid using lipidomic and proteomic methods. *Disease Markers* **22**, 39–64.

45. Yuan, X., and Desiderio, D.M. (2005) Proteomics analysis of human cerebrospinal fluid. *Journal of Chromatography B* **815**, 179–189.

46. Yuan, X., and Desiderio, D.M. (2003) Proteomics analysis of phosphotyrosyl-proteins in human lumbar cerebrospinal fluid. *Journal of Proteome Research* **2**, 476–487.

47. Khwaja, F.W., *et al.* (2006) Proteomic analysis of cerebrospinal fluid discriminates malignant and nonmalignant disease of the central nervous system and identifies specific protein markers. *Proteomics* **6**, 6277–6287.

48. Laspiur, J.P., *et al.* (2007) CSF proteomic fingerprints for HIV-associated cognitive impairment. *Journal of Neuroimmunology* **192**, 157–170.

49. Rozek, W., *et al.* (2007) Cerebrospinal fluid proteomic profiling of HIV-1-infected patients with cognitive impairment. *Journal of Proteome Research* **6**, 4189–4199.

50. Anderson, N.L., and Anderson, N.G. (2002) The human plasma proteome: history, character, and diagnostic prospects. *Molecular and Cellular Proteomics* **1**, 845–867.

51. Luciano-Montalvo, C., Ciborowski, P., Duan, F., Gendelman, H.E., and Melendez, L.M. (2008) Proteomic analyses associate cystatin B with restricted HIV-1 replication in placental macrophages. *Placenta* **29**, 1016–1023.

52. Zhang, L., *et al.* (2010) Proteomic analysis of PBMCs: characterization of potential HIV-associated proteins. *Proteome Science* **8**, 12 doi:10.1186/1477-5956-8-12.

53. Kim, S.S., *et al.* (2007) Different isoforms of apolipoprotein AI present heterologous post-translational expression in HIV infected patients. *Journal of Proteome Research* **6**, 180–184.

54. Rozek, W., Horning, J., Anderson, J., and Ciborowski, P. (2008) Sera proteomic biomarker profiling in HIVâ 1 infected subjects with

cognitive impairment. *Proteomicsâ Clinical Applications* **2**, 1498–1507.

55. Shetty, V., *et al.* (2011) Investigation of plasma biomarkers in HIV-1/HCV mono-and coinfected individuals by multiplex iTRAQ quantitative proteomics. *OMICS: A Journal of Integrative Biology* **15**, 705–717.
56. Cocchi, F., *et al.* (1995) Identification of RANTES, MIP-1a, and MIP-1b as the major HIV-suppressive factors produced by $CD8^+$ T cells. *Science* **270**, 1811–1815.
57. Kaul, R., *et al.* (2000) HIV-1-specific mucosal $CD8^+$ lymphocyte responses in the cervix of HIV-1-resistant prostitutes in Nairobi. *The Journal of Immunology* **164**, 1602–1611.
58. Kaul, R., *et al.* (1999) HIV-1-specific mucosal IgA in a cohort of HIV-1-resistant Kenyan sex workers. *AIDS* **13**, 23–29.
59. Iqbal, S.M., *et al.* (2005) Elevated T cell counts and RANTES expression in the genital mucosa of HIV-1 resistant Kenyan commercial sex workers. *Journal of Infectious Diseases* **192**, 728–738.
60. Valore, E.V., Park, C.H., Igreti, S.L., and Ganz, T. (2002) Antimicrobial components of vaginal fluid. *American Journal of Obstetrics and Gynecology* **187**, 561–568.
61. Burgener, A., *et al.* (2008) Identification of differentially expressed proteins in the cervical mucosa of HIV-1-resistant sex workers. *Journal of Proteome Research* **7**, 4446–4454.
62. Meléndez, L.M., Colon, K., Rivera, L., Rodriguez-Franco, E., and Toro-Nieves, D. (2011) Proteomic analysis of HIV-infected macrophages. *Journal of Neuroimmune Pharmacology*, **6**, 89–106.
63. Wu, W.W., Wang, G., Baek, S.J., and Shen, R.F. (2006) Comparative study of three proteomic quantitative methods, DIGE, cICAT, and iTRAQ, using 2D gel-or LC-MALDI TOF/TOF. *Journal of Proteome Research* **5**, 651–658.
64. Kolkman, A., Dirksen, E.H.C., Slijper, M., and Heck, A.J.R. (2005) Double standards in quantitative proteomics. *Molecular and Cellular Proteomics* **4**, 255–266.

Index